Land

ALSO BY SIMON WINCHESTER

A Crack in the Edge of the World

The Man Who Loved China

Atlantic

Pacific

West Coast: Bering to Baja

East Coast: Arctic to Tropic

Skulls

The Alice Behind Wonderland

The Men Who United the States

When the Earth Shakes

When the Sky Breaks

Oxford

The Perfectionists

The End of the River

Mississippi: Headwaters and Heartland
to Delta and Gulf

The Man with the Electrified Brain

Land

How the Hunger for Ownership
Shaped the Modern World

Simon Winchester

HARPER LARGE PRINT

An Imprint of HarperCollinsPublishers

HarperCollins books may be purchased for educational, business, or sales promotional use. For information, please e-mail the Special Markets Department at SPsales@harpercollins.com.

FIRST HARPER LARGE PRINT EDITION

ISBN: 978-0-06-306301-3

Library of Congress Cataloging-in-Publication Data is available upon request.

20 21 22 23 24   LSC   10 9 8 7 6 5 4 3 2 1

DEDICATED TO CHIEF STANDING BEAR

In 1879 the U.S. government declared this
Ponca chief to be a "person" under the law.

But they still took away his lands.

The first person who, having enclosed a plot of land, took it into his head to say this is mine, and found people simple enough to believe him, was the true founder of civil society. What crimes, wars, murders, what miseries and horrors would the human race have been spared, had someone pulled up the stakes or filled in the ditch and cried out to his fellow men: "Do not listen to this imposter. You are lost if you forget that the fruits of the earth belong to all and the earth to no one!"

—JEAN-JACQUES ROUSSEAU,
Discourse on Inequality (1755)

Contents

Part V: Annals of Restoration | 497

List of Illustrations

UNLESS OTHERWISE NOTED, ALL IMAGES
ARE IN THE PUBLIC DOMAIN OR ARE COURTESY
OF THE AUTHOR.

Prologue
Uncommon Ground

It is a comfortable feeling to know that you stand on your own ground. Land is about the only thing that can't fly away.

—ANTHONY TROLLOPE,
The Last Chronicle of Barset (1867)

A Caveat

With the world's sea level rising fast, the assumption that land is *the only thing that can't fly away,* or *the only thing that lasts,* is for the first time now shown to be demonstrably false. The belief in land's limitless stability has informed humankind's approach to the possession of the world's surface for centuries past, as the following pages illustrate. But now a profound change is coming.

The future is a foreign country: they will do things differently there.

1
Transaction

On a warm midsummer's evening just before the end of the last century, in a book-lined lawyers' office in the pretty town of Kent, Connecticut, I handed over a check for a moderate sum in dollars to a second-generation Sicilian American, a plumber named Cesare, who lived in the Bronx but who had driven up into the lush New England countryside especially for the brief formalities of this day. The all-too-complicated rituals of what in real estate parlance is called a *closing* were familiar to the lawyers, less so to me. In exchange for the check—a cashier's check, certified by the bank to be as good as cash; the lawyers had insisted; and to me it indeed felt like cash, painfully disbursed, and representing some years of scrupulous saving on my part—I was handed by Cesare's stone-faced attorney an engraved, embossed, and rather elegant piece of what looked like parchment, a document formally known as a deed.

This credential indicated that, with the agreed funds having been offered and accepted, I was now the unambiguous, undisputed, and indisputable owner of a tract, formerly owned by Cesare, of 123¼ acres of forested and rocky mountainside, located in the hamlet of Wassaic, in the village of Amenia, the town of Dover, the county of Dutchess, in the state of New York. The deed gave me title to the tract. I was now legally *entitled* to possess it. I was its owner. I could now occupy it, exclusively.

I had just purchased a piece of the United States of America. A morsel of the continent now belonged, exclusively, to me. Since the total land surface of the planet amounts to some 36,652,096,000 acres—let us say 37 billion acres—I could declare, on walking out into the sunshine on that July evening, that my 123 acres, a little over three-billionths of its extent, were now mine, and mine alone.

It was the first time in my life that I had ever done such a thing, had ever come to own a piece of real property, anywhere. Personal property, yes: I had owned cars, computers, dishwashers, books, fountain pens. But real property, and its kissing-cousin real estate:* this was a first. A first for me, and more or less the first time for anyone in my family. Back home in England my parents, whose lives had been mostly spent in respectable impoverishment, had in their later years managed to buy a small cottage in the English Midland county of Rutland. Since their house came with a tiny postage stamp of a garden, with a lawn and some shrubs and a dribbling water feature, it can technically be said that, once they had discharged their mortgage, they did indeed own real estate, albeit only a vanishingly small parcel of it. It would be stretching things, straining credulity, to hazard a description of them as landowners.

It was much the same for my grandparents, whose circumstances were similarly straitened. Both the paternal-side couple who had lived in England, as well as those members of my mother's rather more complicated side and who came from Belgium, had had nothing of the sort. For them, as for many people, accommodation and

---

* There is a subtle difference between the two terms: while *real estate* has come to mean the tract and any improvements made to it—a house, for instance—the term *real property* signifies in addition the so-called Bundle of Rights that can come with simple ownership: the rights, for example, to occupy, to sell, to mine, to clear away timber, to prevent trespass, to exclude others. Further explanation later.

shelter relied on the whim and commercial acumen of a series of landlords, who as the name suggested had themselves owned the acreage on which my various grandparents had settled. I was given to understand that all the generations of my ancestors before them were un-burdened by ownership also, had never owned an acre or a hectare, a cordel, a fardel, or a virgate,* nor any fraction thereof. All of which made my receipt of that Dutchess County title, a document of such letterpressed and engraved magnificence that I would gaze at it en-raptured for hours, both historic and precious.

Even though my property was neither particularly costly nor likely ever to be valuable, nor in truth was even very useful, my ownership of it had a powerful personal symbolism. A decade or so after the trans-action I traveled to Boston, to the afterdeck of an old sailing warship, where I swore an oath before a fed-eral judge and became, in a brief but moving cere-

---

*   A reasonably complete list of familiar, foreign, obscure, ob-solete, and rare units of areal measurement will appear in the glossary. Regrettably the list will not be able to include such non-areal measures as the *glean*, which applies to English herrings (1 glean being 25 fish), nor the German unit for 15, the *mandel*. It does include the *ping* and the *pyong*, respectively Taiwanese and Korean units of area, as well as the *dunam*, traditionally employed for measuring fields in the Balkans.

mony, an American citizen. For many years after that life-altering event I drew considerable comfort and satisfaction from knowing that I had become fully invested, and with the square footage of mountainside under my ownership quite literally so, in the future of my new country.

I would walk the forest—my forest now!—as often as I could. I would puff and pant my way up the escarpment, following a vague and almost vanishingly overgrown track through the woods and which led up from an ancient loggers' landing. After a quarter mile or so, I would turn off left into the deeper timberland, after which I would frequently get somewhat confused, disoriented, and even a little lost. I remembered from Boy Scouts days that moss tends to grow on the north side of the trees and from school physics days that streams tended to flow downhill, and reckoned therefore that I would always be able to find my way out and back to civilization. Moreover, it turned out that, slicing across this tract of mine in a perfect die-straight line of cleared underbrush, and with permission seemingly granted by me under grandfathering laws in which I played no apparent part, and with warning notices posted every few hundred yards in red and white, there was, in a buried, invisible, and supposedly atomic-attack-hardened cylinder of concrete, a secure and once secret communications cable that had connected the Eisenhower-era White House with a strategic nuclear bomber base somewhere up in Maine. If I stumbled across this, then I knew I would find my way out—at least to Washington, or to Maine.

But otherwise, once deep in the woods all the world soon faded, the forest became almost primeval in its quiet and detachment. Somewhere perhaps, and not too far away, maybe, there would be a rain-softened and moss-covered split-rail fence of cedar, perhaps an old wall of tumbled stones, and by chance a cairn and a chiseled mark or two left on a rock by some long-ago survey team. But generally, deep in these dark woods' darker middle, there was no other clear sign of human intervention or activity, few clues for the passing stranger that mankind was ever here at all. There were just trees and ferns and soil and birds and the leavings of deer and rabbits and raccoons and bears, and overhead, glimpsed blue and silent through the crowns of leaves, the presiding vastness of what the poet John Clare, two centuries before, had called *the circling sky.*

On that first untutored glance the surface here appears today as it always was, wild, unchanged, the result of geology and weather and heat and pressure and time. Most of all, by near endless extents of time, which all around the world, and not just here, have rendered the once malleable and plastic planetary crust into something colder, harder, and workable, and on which in many places, such as this forest, life of all varieties has come slowly to exist.

Speakers of the English language have long called this exposed and rendered surface by a word that has been part of our vocabulary for longer than almost any other. It is the word *land*, a word formally denoting that exposed portion of the planet which is higher than and is fundamentally physically different from—and by happenstance is also somewhat less extensive than—that part which is today covered by water and which since the thirteenth century we have called the sea. *Land* is an originally Germanic word that has been current in English since the tenth century, denoting since then the solid surface of the planet that is found generally lying above sea level.* What strikes many as ironic is that we have long called our planet the Earth,

* But not always: the shores of the Dead Sea in Palestine, the Turfan Depression in China, and Death Valley in California are all, as a result of geological accident, substantially below mean sea level. Thirty-three countries have areas which are in theory liable to marine inundation—a number that will increase as the oceans rise and low-lying coastal strips and small islands are drowned, thanks to the warming of the seas and the melting of continental ice sheets.

when—and this is of course especially noticeable when our blue and green spheroid is seen from outer space—it manifestly should more properly be called the Ocean. The Earth is more sea than land, and by a long chalk.

Some may wonder why the word *land* came into common usage so much earlier the word *sea*.* There is a reason, and it points to two fundamental differences, other than the obvious physical dissimilarity between earth and water, and which so markedly separates the two. One difference is that while the sea generally looks—and indeed is—much the same everywhere around the planet, varying only slightly in apparent color and warmth and salinity—the land varies hugely in aspect from place to place, and often does so in close proximity—there are mountains here, valleys there; there is desert or glacier, swamp or meadow, the surface is undulating or jagged, fertile or barren, wooded or grassy, its features hot and dry, or bitter cold and edged with ice. Variation of landscape is a basic feature of land, and is something to which inhabitants are sensitive, and of which they presumably always have been profoundly aware.

---

* The OED in fact suggests that the plural *seas* was first used earlier still—in AD 825, compared with *land* in 900 and *sea* proper and singular in 1275.

By contrast—and this is the second fundamental difference—land and landscape are the near exclusive domain of air-breathing mammals—and most especially those that can speak, read, and write. And even before humankind first encountered the sea, humans would have been aware, and just because of its endless variety, of the landscape in which they were placed. They would have noticed, and noted—with the result that land and its vast spectrum of forms would have come into their vocabulary with more facility—its sheer variety made them more aware of it. Had there been note-taking creatures living in the early sea, this watery medium in which they lived would have initially been invisible to them, and the vocabulary needed to describe it would necessarily have been rather limited in extent.

2
Foundation

Like all land and all landscapes, everywhere, mine has a story to tell. It has not always been owned. Nor, for that matter, has it always been land.

The exposed bedrock in this particular part of North America is extremely ancient. Those benches and ledges of hard and lichen-draped rock that I can see poking out from the thin skin of forest soil, from among the dead leaves and mosses on the floor—as well as the clues offered by the boulders piled in the surrounding walls—are considerably more ancient than in most of the rest of the country. They are very more varied, too. In these square miles of rural New York State more than one hundred different rock formations have been described by geologists who, for decades past, have tramped enthralled across these hills and along the river valleys, exploring with their hammers and magnifying glasses and bottles of acid, with their compasses and clinometers and their all-too-keen eyes.

The story that these New England rocks tell is one of hundreds of millions of years of geological turmoil executed on a titanic scale. It is a tortured and spectacular history that begins with volcanic land formation, then is given over to eons of sudden fracturing, splitting, compressing, heating, pummeling, twisting, folding, and breaking, followed by millions more years of inundations by tropical seas, with the much altered first rocks now covered by thousands of feet of deposited new materials, after which and all in good time there were long periods of slow upward thrusting to the surface, further long drownings, and then immense and violent collisions with other bodies of rock that were either broadly similar to or else very different from the first formed fragments of the volcanic crust.

All of this activity, and all of these processes, helped to forge and deposit what are today's New England rocks—what is today's New England land—and it forced these rocks together in the improbable mash-up of granites and limestones and shales and sandstones and basalts that now extend from the Hudson River valley to the Atlantic Ocean. This part of what in today's world is located well inside the Northern Hemisphere lies essentially halfway between the North Pole and the equator. My three one-billionths possession of the total land-and-sea planetary surface stands at 41.8 degrees north latitude—halfway to the Pole would be 45 degrees, so it is a little to the south of the center point. It is a stark reminder of the topsy-turvy nature of the prior workings of the planet that all of the building that put my land where it is today actually took place in the early Southern Hemisphere, close to the old South Pole, thousands of miles away from where the land has ended up today. The beginnings of my tract of land were both long ago and far away.

Very long ago. Basically, the hill up which I came to like to trudge was once very close to—I like to think it actually was—the extreme edge of Laurentia, one of the world's earliest continents. Laurentia was not the oldest of the continents, not by a long chalk—Kenorland, Nuna, Ur are the names given to even older such bodies, with the calculated oldest of them all, Vaalbara, rising from the hot primeval ocean about 3.6 billion years ago, only a billion years, a mere temporal bagatelle, after the world itself first formed.

Laurentia is very much younger than this, and is now known to have been just a fragment of a gigantic supercontinent that Russian geophysicists, specialists in the mind-bending subtleties of tectonic forensics, named Rodinia, the mother ship of the modern world. This huge body of proto-real-estate was created about a billion years ago and then promptly—after just 250 million years, a blink of an eye in the cosmic scheme of things—broke into two, giving us first Gondwana, which remained in the south and went on to spawn progeny that included Australia, Antarctica, and India; and then the second, my piece of land's particular birth mother, Laurentia, which then drifted with due majesty, up and into the Northern Hemisphere.

After which matters unfolded much more rapidly, with more and more varied kinds of geophysical and biological events compressed into less time. Life began, with the appearance of sea creatures initially, and then in time the great arrays of plants and motile beasts that flourished on the land, in those periods when land itself appeared. A new body of water, the Iapetus Ocean, began its existence off the eastern shore of Laurentia, and limestones, with life that eventually became fossils and which can be found once the limestone became marble, were laid down within it, on the continental shelf. Volcanoes spurted up from under the sea and caused the formation of arrays of newer igneous rocks that adhered to the limestones and the beach sands and the old rocks of the Laurentian coast; and then there were mountain-building episodes, first those of 450 million years ago, which helped form parts of Maine and Nova Scotia, and then, more dramatically still, the moment when the prodigal child Gondwana traipsed back up north to reunite violently with Laurentia once again to form a new great continent, Pangaea, and in the process, to create the Appalachian Mountains and the Berkshires of which my tiny piece of land is a barely significant peripheral component.

Pangaea then duly broke apart, as all supercontinents have been wont to do, ultimately leaving the world looking more and more as it does today—but not fixed today either, with the Atlantic Ocean now separating the Americas from Europe and Africa, widening every year, centimeter by centimeter, to produce in due course yet further changes in the geography of the world.

It is best to think of the landscape as static now, insofar as the human clock runs so much more frantically than does the somnolent passage of geologic time. And so here, a legatee of all this history and turmoil, all of this sturm und drang, now set and seemingly motionless high up on the eastern side of North America on a block of uplifted land between the recently ice-sculpted valleys of the Hudson and the Housatonic Rivers—the one river named for an English explorer who in 1609 was exploring and mapping it, working initially for the Dutch; the other a Mohican Indian word for "the river over the mountain"—lay the tiny piece of real estate that would eventually become mine.

My piece of land has lain for most of its existence in the climatically temperate zone, so far as botany and zoology were concerned, and it was soon to be covered with a dense carpet of trees, shrubs, and grasses, and populated by a fair variety of beasts and birds. The countryside around was, as Henry Hudson wrote, "as pleasant with grass and flowers and goodly trees as ever they had seen, and very sweet smells came from them."

In all likelihood the lands that Hudson glimpsed and then visited, as he ventured out from his river flotilla to meet and trade with the local natives, were in the valley and the river's floodplain, terrain with thick rich soils and ample cultivation. Up in the hills, where the weather was more harsh and the soils more thin, the vegetation would have been less florid, less exotic, the population scarce, the agricultural methods of the natives hardscrabble. Forest dominated then, as it does today. My own land nowadays has stands of eastern hemlock and pignut hickory, tulip poplar and white pine, beech and birch, conifers and ash, cherry and witch hazel and two kinds of maple, one with sap to be boiled for sugaring, the other tiger-striped and used today in the making of furniture. Apple trees can be found once in a while, though none on my property, so far as I know.

No doubt all these trees would have been there in Henry Hudson's time, and some for long before. There used to be American chestnut trees—a blight has ravaged them in recent years—and it was customary to plant them as corner markers to define the edges of a plot of land. Chestnut was once the dominant wood for building barns and houses, but not since the trees were forced into near extinction.

The soil is too thin to support massive species—no great oaks or elms, no trees tall enough and straight enough for repairing or replacing the masts of big sailing ships—so often a motive for a ship's master arriving in a foreign land after long sea crossing. Modern soil science classifies my forest floor as having a mix of what are classified as the Charlton and Chatfield types, soils derived from the residue of retreating glaciers, colored brown and in texture loamy with gravel in places, sand in others, and being on average less than a foot thick. Solid and unweathered bedrock—granites, schists, gneisses, and beds of marble, all metamorphosed by the millions of years of violence, heat, and pressure—lies about two feet below the surface, and everywhere exposes itself in ridges and shelves, on which, in the languorous heat of summertime, timber rattlesnakes are known to laze. A local herpetologist later told me, with a glint of specimen envy, that my land is uniquely and legendarily rich in endangered rattlers. In the twenty years I have been the owner I have seen just one, coiled up on the far side of a wooden bridge across which I was planning to run. That particular day I decided to wade through the brook instead, and for many months afterward gave the bridge a wide berth—even though common sense told me the snake was likely to be far more scared of my presence than I should have been of its.

There is much wildlife, with the impenetrable stands of mountain laurel providing cover for the white-tailed deer that are still much hunted in the autumn, as well as black bear, coyotes, raccoons, opossums, and, on occasions, lynx and wildcats. There used to be wolves and panthers, and even today people claim to see occasional mountain lions. The little stream that slides under the wooden bridge is often dammed by a family of beavers, competing amiably for territory along stretches of the stream with groups of otters. Wild turkeys are everywhere, huge families of birds of all sizes, and which march through the woods like soldiers on maneuvers.

There are songbirds too—some rare, like the hermit thrush and the eastern pewee—and frogs and salamanders and monarch butterflies, as well as, cunningly lying in wait with tiny invisible strands that cling to the clothing of passersby, all too many deer ticks.

3
Population

And then, into this generally congenial mix of flora and fauna, and into a climate that generally favored mammalian existence, stepped the first people.

These were of course fully formed and fully civilized people—no early hominids ever populated the Americas, no prehumans, no Neanderthals. These were *Homo sapiens*, pure and simple. True, dotted across New England there are a goodly number of Paleo-Indian sites, suggesting that as much as 13,000 years ago, rather less sophisticated hunter-gatherers roamed their way through these woods, bringing down their prey with spears tipped with distinctly shaped points, shapes through which the various peoples are now archaeologically classified. But the great majority of those natives who descended a little later to settle on these particular hills and valleys were well-advanced tribal families, descendants of those Asian pioneers who had trekked across the Beringian land bridge—or had come by boat from what is modern-day Japan—to spread out across an otherwise uninhabited—but in most places, eminently habitable—continent. Those who came to this corner of New England were speakers of what are called Algic or Algonquian languages,* and they were members of the Mohican tribe—related to the Lenape people to their south, and sworn enemies of the Iroquois and the Mohawk to their north and west.

Hudson described the Mohicans, whom he and his crew termed the River Indians, as a gentle people, and though there were skirmishes, some lethal, and probably more the fault of the invaders than of the generally peaceable and innocent natives, the first contact was generally civil, the visitors impressed. Gifts were exchanged: the Indians offered corn, pumpkins, and

---

* Recent research by Russian linguists has suggested a connection between Algonquian and the isolated language spoken by the Nivkhi peoples of Sakhalin Island, north of the Japanese island of Hokkaido. Since other inquiries have suggested, thanks to the discovery of a relic boat, a possible trans-Pacific Hokkaido connection with North America, this linguistic link, though seemingly fanciful, is perhaps not wholly improbable.

tobacco, the sailors gave mirrors, bells, beads, hatchets, and knives. Hudson's men were welcomed ashore, and found along the river a series of settled and contented communities. The Mohicans lived in villages of twenty or so longhouses; as well as being competent hunters and trappers—their winter clothes were heavy with fur—they were also very evidently skilled agriculturists, growing corn in large fields beyond the village pale, as well as harvesting squash, sunflowers, beans, and all manner of berries that grew wild and profusely in this elysian corner of the world. There were sturgeon, lamprey, bass, and eels in the rivers; the hills were thick with deer, moose, and elk. Even in the harshest winters these stable and sensible people—thousands of them, their settlements visible everywhere—survived and prospered, seldom going hungry or unclothed. They seemed to belong to a happy and healthy and well-organized society and, at least at first, were more than amenable to the arrival of strangers.

And so boatloads of strangers—the first of them from Holland, backed by the Amsterdam-based Dutch West India Company—began to arrive. After Hudson's reconnaissances on their behalf, the Dutch port of New Amsterdam was established on Manhattan's southern tip in 1625. In due course expeditions from England and then France began to trickle in too, starting in earnest in the middle of the seventeenth century—Massachusetts Bay, Virginia, Quebec—territorial claims that would be the spur to many a battle in the years to come. The newcomers were all white men—joined eventually by white women—and as with many traders, settlers, and all European colonists, everywhere, they were sustained by an unshakably confident belief that they, simply for being white men, were superior in the world to all else. And at a stroke, and with the striking of such an attitude, so the comparative serenity of the region changed, and everything, from racial harmony to personal health, started to deteriorate.

The serenity of the Mohicans suffered, terminally. The villagers first began to fall fatally ill—victims of smallpox, measles, influenza, all outsider-borne ailments to which they had no natural immunity. And those who survived began to be ordered to abandon their lands and their possessions, and leave. To leave countryside that they had occupied and farmed for thousands of years—and ordered to do so by white-skinned visitors who had no knowledge of the land and its needs, and who regarded it only for its potential for reward. The area was ideal for colonization, said the European *arrivistes*: the natives, now seen more as wildlife than as brothers, more kine than kin, could go elsewhere.

And due to the pitiless twin effects of illness and expulsion, so the Mohican people in this corner of the New World withered steadily away—as James Fenimore Cooper chronicled so hauntingly in *The Last of the Mohicans*, published in 1826 but set in 1757, and a perennial favorite of Hollywood. Despite reams of promises made and entreaties offered, they eventually trooped off, unhappily, one of the lesser-known Trails of Tears, to new homes in the very different territories of Wisconsin and up north in Canadian Iroquois land.

A subordinate tribe of the Mohicans with the tongue-twisting name of the Schaghticoke lived on, however, settling on the eastern flank of the mountain of which my land forms a 123-acre part. Theirs is a luckless story, too. Their early-eighteenth-century leader, remembered still in Connecticut simply as Chief Squantz, declined to sell his land, amounting to a modest 2000 acres, to a group of white men, known as the Proprietors, who were searching the Housatonic River valley for a place to construct a colonial community they would later call New Fairfield. The chief died, stubborn to the end, and his offspring turned out to be equally implacable. But just four years after their refusal the land was taken away anyway, and the tribe was given three hundred dollars and told how lucky they were to get anything. Racist contempt was a characteristic of all too many of the colonists, with Native American chagrin and disappointment its twin.

The relict Schaghticoke did not entirely leave, however, but clung on in the naïve hope that one day, treaties would be honored and promises kept. But they never were; and today the rump of the tribe, with a few hundred others scattered wide and bickering among themselves, are reduced to living in a cluster of shacks on the flood-prone banks of the Housatonic, constantly in and out of court, pressing in vain their claims for land long ago given to a local school and the state power company, and likely never to be Indian-owned again. The dirt road that passes their tribal headquarters—I use it to reach my small tract, if I am driving up from New York—is a miserable, ill-maintained affair, and it floods deep when the river rises in spring. Since they are so few, the Schaghticoke have recently been struck off the formal list of federally recognized tribes—there are 360-odd, and my neighbor tribe was until lately on the roll between the Sault Ste. Marie tribe of the Michigan Chippewa, and the Scotts Valley ditto of the California Pomo. Now they are almost all gone, humiliatingly known more by their absence—and with the local Mohican far away by the Great Lakes for the last two hundred years, so my mountainside has been ripe for western settlement and colonization, which has been its fate ever since.

The newcomers, eager legally to secure the taking of the abandoned native lands, introduced one formality that their Indian predecessors had never known: the title deed. Such a document, much like that which I would be handed three centuries later, soon became an essential for demonstrating that one was actually the rightful owner of a piece of real estate. One of the earliest such New England title deeds was written in 1664, when a Dutchman named Willem Hoffmeyer bought three small islands in the Hudson, recording the purchase from three Mohican chiefs. His handwriting on the deed, and that of his clerk, is florid, legible, and adorned with extravagant curlicues. The Indians were naturally unversed in the writing or reading of English, and their signatures on the deed are, respectively, a line drawing of a bull, a turtle, and a field of corn. The simplicity of their part of the document has a poignancy all of its own, when it is considered how utterly dispossessed these three Mohican men would likely be, probably before the ink on the agreement was dry.

Other, more prominent Dutchmen, with questionable timing and capricious loyalties, then played a signal part in the development of the lands I would

An early title deed, showing the supposed transfer of Hudson Valley land from Native American to colonial ownership.

one day purchase. The best-known name, commemorated in towns and manor houses and railway stations to this day, is Philipse, a crude anglicization of Flypsen, under which name the patriarch of the family, Frederick, arrived in the Dutch possessions of the New Netherland in 1653, when he was just twenty-seven. He was a typical young merchant venturer of the time, aggressive, eager, and canny—and he made his first fortune selling square-cut nails to fellow settlers who were building houses. He then turned to hospitality, and built and bought a number of taverns to sell these same men strong drink to refresh them after their day of construction. Once financially secure, by about 1672, when he was forty-six, he commenced a land-buying spree, acquiring his first tract from another, rather less successful Dutchman who had bought (for strings of wampum beads, still the Dutch colonial currency and one recognized and accepted by the Mohican and Lenape Indians) these tracts from the local natives.

Two years after his first purchase, of some 80 square miles of Hudson Valley land to the immediate north of New Amsterdam, the Dutch capitulated in their latest war with the British. England promptly took over—changing the city's name to New York—and after Philipse had most expediently switched his loyalty to the incoming British, he was made lord of the manor of his already very considerable acreage, which would become what it still is today, New York's Westchester County.

Philipse had a formidable family, both in its quantity and quality*—and, despite the paterfamilias being a Dutchman (albeit of Czech origin), all of the offspring displayed, once it was politically prudent to do so, an unwavering loyalty to the British Crown. Frederick's son Adolphus (the second of eleven children) was quite as land-obsessed as his father, and in 1697 he bought some 250 further square miles of territory from a passel of Dutch traders, and consolidated his purchases into one huge bloc that became known as the Philipse Patent, which was, thanks to Adolphus's aforementioned loyalty, royally sanctioned from London. This enormous tract of rich and fertile land had in 1683 become Dutchess County, named by the English for the soon-to-be-King James II's Italian-born consort, the Duchess Mary of Modena, but using the archaic spelling of her title, Dutchess. More than a century later, once the local population numbers began to swell, the

---

* Among his many descendants was John Jay, the country's first chief justice.

local colonial burcaucrats thought it was becoming too challenging to administer, with the result that the more southerly acreage was eventually cordoned off, meta- phorically, into the quite separate Putnam County.

Dutchess, the more northerly portion of the Patent,* was thus for the first hundred years of its existence a privately owned fiefdom within the Royal Colony of New York—the Philipse family essentially owning by kingly courtesy and historic right the real estate; and the Crown back in England owning by divine munifi- cence and fine-tuned hubris the real property, enjoy- ing the fundamental seat of ownership. This effectively meant that until 1776—when of course most Americans threw off their colonial yoke and won independence, and the Philipse ownership was ended and their lands confiscated by the new-made state—anyone wishing to

---

* Arguments over land and surveys in the late seventeenth century between the two neighbor colonies of New York and Connecticut resulted in the curious bizarreries of their current shapes: the strange Connecticut panhandle down south, now stiff with millionaires' mansions, and the more bizarre sixty-mile- long and two-mile-wide Oblong—its official name—granted to New York on its eastern edge. My land lies very much in the middle of the Oblong, and until 1713, was part of Connecticut Colony. Had it remained so, my property taxes would have been much reduced.

live and work in this corner of Britain's new possession could do so only as a tenant. The temper of the times was very clear: tenancy was welcome, but the notion that any individual could own, could have title to the land, was both impertinent and absurd.

4

Exploitation

The economic development of the region was as a result lackluster, at best. The incentive for improvement, a virtue common to owners rather than to tenants, is systemically lacking when absentees hold the title—an axiom that goes to the heart of the very principle of landownership. Yet here in and around the New York Oblong, the colonial economy, lackluster though it may have been, was far from negligible. This was largely due to a happy accident of geology—the presence of an abundance of a fair-to-middling quality material that was much needed both at home and abroad: iron ore.

Caught up in the tortured rocks that were twisted and hammered by the Taconic Orogeny, 450 million years ago, there were and still are huge deposits of fair-quality hydrous iron oxide. Insignificant by the standards of Minnesota's immense mother lodes, the bodies of ore that were found in and around the Oblong were sufficient in the eighteenth century to tempt English ironmasters to cross the ocean, panting for business. They had the expertise and now they had the lure: they could mine the ore, smelt it in charcoal-fired blast furnaces, cast it into cannon and anchors—and nails for the building of houses—and in the process make themselves a fortune.

The first iron was found near Salisbury, Connecticut, in 1731: the first blast furnace, a monster of firebrick and cement thirty feet tall, was built soon thereafter. Within a decade the region had become the principal iron-making region of America. Its first products—ploughshares to penknives, hammers to nails, guns to teaspoons, and ships' anchors, taken by horse-hauled wagons to the wharves at Poughkeepsie and thence down the Hudson to New York—were used by the colonists initially; but then the American revolutionaries captured the furnaces and turned the smelters and the forges to the manufacture of their own necessities—cannon and flintlocks and musketry—for warfare and for struggle. Then the new independent government of the United States took control, and the business of arms manufacture—and the making of more delicate machine parts for the inventions of the Industrial Revolution, which had been born, by happy coincidence, in the very same year as the Declaration of Independence*—became ordered and dependable and imbued with what would become an awful power.

Four basic ingredients were needed for the making of iron goods: iron ore, lime, water, and fire. The ore was geologically abundant in and around far western Connecticut and eastern New York; the lime was quarried from the Stockbridge marble deposits close by; there was water everywhere, in lakes and rivers and rivulets and brooks. And the fire was made from charcoal— which could be made in fire pits excavated in the forests on the hillsides in places like Wassaic. The men and women who came to Wassaic and who first settled near my land in the mid-1700s came there principally to make charcoal. They made huts for themselves; they cut and felled the trees; working in concert with one another they hauled up enormous stone-compassed charcoal pits, ten yards in diameter, and in these they

---

* James Watt's first properly efficient steam engine, its much sought-after efficiency the direct result of the mechanical perfection of its main cylinder which was created for Watt in North Wales by the technologically adept ironmaster John Wilkinson, had its first trial run on May 4, 1776—inaugurating the new era of modern mechanical world.

stacked the logs and slow-cooked them to make the high-temperature fuel that was needed by the foundries down in the valleys. The circular pits were in fact called coal pits locally, even though there was no coal for miles, and charcoal was the man-made substitute. It required two tons of ore and limestone—and 150 bushels of charcoal, most of it made in the woodlands that I and others now possess—to make one gross ton of iron.

All of these coal pit workers were rightly pioneers. As to whether any or many of them claimed or wished for actual ownership of the land they farmed, the records of that long ago do not tell. All one can surmise with certainty is that up to a certain well-remembered moment in late-eighteenth-century history, and possibly before that, possession of the land in what was first New Netherland, and then New York, was claimed by the titular head of the nation that been the colonizing power: until the Dutch surrendered their colony in 1664, there were a succession of *stadtholders*—Maurice of Orange perhaps the best known—after which it was England's turn, and the titular local landowner was King Charles II, and then James II after him, and their successors until the hapless George III, who in 1776 famously lost it all.

In reality the charcoal makers up in the forests, whether in Dutch or English times, were originally tenants of the Philipse family, given the existence of the Philipse Patent. But that all changed, not in 1776 when independence was proclaimed, but four years later, by which time the government had organized itself well enough to announce that the Philipse lands and property had all been confiscated—the family having been, rather ostentatiously and vocally, loyal to the British Crown. Their tenants were in consequence, and from March 10, 1780, quite free actually to buy their farms and the surrounding lands. Impertinence and absurdity were cast aside. The locals could at last own a portion of the new United States and be fully invested in the nation's future.

5

Demarcation, Eviction, Possession

The first title deeds were written; the Philipse estate was divided up by a new government body called the Commission of Forfeiture, and some two hundred new parcels, some of them existing farms and smallholdings, others newly surveyed and deemed ripe for building, were created.

And, however brutally, the land had essentially been emptied for their use. By now almost all of such Mohicans as had lived here had been turned out—a few hardy settlements of what Henry Hudson had called these "gentle people" did manage to cling on well into the eighteenth century, after independence, but before too much longer the exodus got properly under way, with families moving either to the west, and the far side of the Hudson River; north to a mission in Stockbridge,

Massachusetts; or joining the larger pilgrimage and final exodus across the Mississippi, to Wisconsin. There was the single local exception of the Schaghticoke, whose influence persisted long after Chief Squantz's death in 1724, the legacy of whose stubborn determination to retain lingers even to this day. Otherwise, in what was now postcolonial America, the hills were given over wholly to wild animals and birds, to ferns and shrubs—and a superabundance of wood.

Little more than the vaguest shadows of Native American occupation, centuries-long perhaps, maybe even millennia-long, can be discerned or surmised, even though they would be, should by right and human decency be, the true owners of the land. The first owners, one might say. In recent times some communities in New England have taken to offering—with a brief speech, a short homily, a moment of silence before a public meeting—a token of respect to those Native Americans who came first. This has already been done for years in Australia and New Zealand, countries whose treatment of their own aboriginal predecessors is no less lamentable than America's has been. That the notionally redemptive practice is now spreading slowly up to—and by way of New England now into the public fabric of—the United States can do little but good, even if it comes a century or more too late. If nothing else the pause it brings serves to remind us that when we reach back so far in time, questions about the precise nature of the ownership of land, in the United States and elsewhere, become necessarily ever more vague and the answers less fathomable than they are today.

My own land, possibly at first no more than a hunting ground for the Native Americans, was for the white settlers typical of what they expected, if topographically somewhat harsher than tracts at a lower altitude. Higher and less hospitable meant, perhaps, that it was cheaper. Then as now, it comprised a north-facing hillside, steep, thickly forested and liberally strewn with glacial erratics. In all likelihood it had never been properly settled, probably had seldom been trekked over other than by barefoot huntsmen. The fact that today there are, except in the deepest recesses of the forest, a scattering of decayed fences and moss-covered walls and boulder piles and, tucked secretly away in the recesses, the pits of the charcoal makers, suggest it was owned and used. Just not lived in, full-time. Not at first.

The land was also mapped in only the most rudimentary manner. That changed swiftly, though: the craft of mapmaking, already instilled and taught by the departing British army cartographers, was keenly adopted in the infant America, and maps and charts and surveys of the individual plots were soon drawn and engraved and lodged—as items of singular beauty, as well as of legal necessity—with the record-keeping authorities.

A map in the Library of Congress published in 1850 shows, where my land appears to be, a farmstead belonging to a family called Wilcox and another to Wilson. But these are low down on the east side of the hill, and there is no evidence of a track leading up to the mountain. The presumption has to be that this mountaintop land—rich with rattlesnakes, it is noted today—was seen as unattractive, and was almost free for the asking. And though some local aristocrats had their eyes on the mountain, as a means simply of augmenting their existing landholdings and showing off the areal dimension of their wealth, they were soon seen off by men of humbler means, who felled trees and built pits and made charcoal, and kept outsiders at bay. It is claimed that groups of freed Black slaves were given parcels of the nearby land after the end of the Civil War; but few long-settled African American families are to be found in these hills today, and if slaves did come, it seems that they left in short order.

I soon found that my land enjoys a border—marked

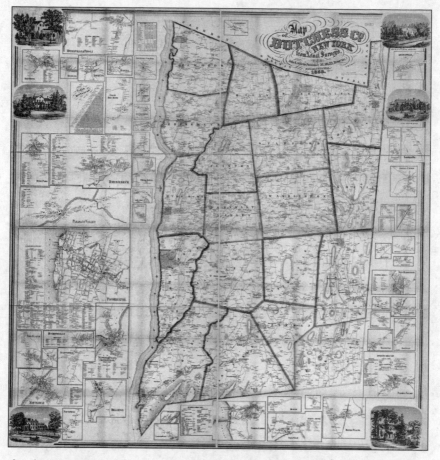

A nineteenth-century survey map of Dutchess County, New York. The land under discussion is in Amenia, in the northeast.

with colored blazes on some taller trees and nothing more—with that of a family called Brasher, the patriarch of which was Rex Brasher, famed locally for drawing birds in the style of, but most would say more competently and artistically than, the mid-nineteenth-century ornithologist John James Audubon. In 1911

Brasher spotted an advertisement in the For Sale section of a New York newspaper: *FARM: One hundred fifty acres, half wood, half brook pasture, good house, barn and buildings. Terms reasonable.* It was in a community of which he had never heard: Wassaic, home to an iron foundry, and the country's first-ever condensed milk factory, run by Bordens. And surrounded by woods full to bursting with birds.

So Rex Brasher bought it, for a song: and twenty years later he published, privately, his twelve-volume *Birds and Trees of North America*, widely regarded today as the greatest and most definitive set of paintings of every major bird species living in the country. His relatives still own the farm: they liked to walk through my woods during the autumn hunting season in bright orange clothing, to warn off hunters. They abhorred—as Rex Brasher did, vehemently—the practice. Wildlife was for painting and recording, they said, not for killing for sport.

But their mission was in vain. Hunters did come, in droves. There is still the wreckage of an old hunters' cabin on one spur of flattish land, leading me to suppose that parties of men with guns arrived from New York City, a hundred miles away, to shoot and carouse happily after a successful day's killing. One owner of whom I came to know was a Sicilian immigrant named Sebastiano Vacirca. From what I gathered Mr. Vacirca, who came in the middle of the twentieth century, was a keen hunter, and since even now the woods are thick with deer one imagines a few trophies, and a few thousand pounds of venison, were taken over the years.

And then, like the ground coming up fast for a parachutist, so the cascade of modern ownerships are recorded, quick-fire—the family of a man named Edward Doll, and then Cesare Luria, who owned the parcel for twenty years and would come up from the Bronx on autumn weekends in the 1970s and 1980s, to chase down the deer and after gralloching them and preserving the antlers for the mantle, to lug the venison in coolers back to his friends and associates on the streets of New York.

And then one July day in 1999 I came along, fascinated by and longing for this land, which had been advertised as for sale itself, and I duly parted with my cashier's check and my hard-earned, sedulously saved moderate pile of cash, and shook hands with Cesare to seal the deal. We began a friendship that still survives, cordially—and with permission to hunt sought annually by him and granted near automatically by me. Coolers of venison are left out for me, not infrequently with a bottle of cognac inside for good measure when the season has been a good one. All this, though, was for later: on this July evening in 1999 we shook hands, and Cesare turned on his heel and headed home, and suddenly I was the owner of his land.

6

Exploration

O nce the real estate closing ceremony was over and done with, I lit out westward down the road from Kent, and ten minutes later, just past the sign for Macedonia Brook State Park, and the granite state-line markers that indicated the beginning of New York—the beginning and eastern edge of the Oblong—and after I had turned down the little dirt road that Rex Brasher had once called Chickadee Road because of the swarms of such creatures in the trees, and after half a mile bumping along the gravel and the potholes, I was there.

I stepped out of the car and into the cool shade of the evening. In a trice I was in my forest, thinking of what exactly I had just done. And wondering: What does my ownership of this land truly signify? What does it mean, to own land? Surely land, I said to myself, is an entity that cannot really be owned, by anyone.

And yet, in legal terms, this tract had become possessed. This tiny morsel of the planetary surface, a barely significant three one-billionths of it at best, has, at various times in its maybe five thousand years of populated existence, been owned or claimed or settled, among others:

by groups or families or communities of Mohican
    Indians,
by groups or families of communities of
    Schaghticoke Indians,
by—at least notionally—three Dutch *stadtholders*
    in the House of Orange-Nassau,
then in a titular sense by a number of English
    monarchs, the Stuarts Charles II, James II,
    Mary II, William III, and Anne, and then after
    the English became British the sad Hanoverian
    trinity of the first three King Georges,
and during this post-Dutch time the land was
    owned or presided over by the patriotic-
    Dutch-turned-English-loyalist Philipse family,
    as Patentees,
then once these were booted unceremoniously
    from their holdings, the land formerly tenanted

by lessees of Philipses came actually to be
owned, in *fee simple* as will be explained later,
by farmers and hunters and charcoal makers
whose names have passed into undocumented
oblivion until it was owned
by a family named Brasher who came to live close
by and then
by a Sicilian immigrant named Vacirca, then
by a German American named Doll after which it
was bought
by a Sicilian American named Cesare Luria—and
finally,
by the current owner—who became transfixedly
fascinated with the notion of landownership,
and of how such a thing could possibly be,
and why so many all across the world seem
to go to such great lengths to acquire, to
mortgage themselves for, to fight for, to steal,
to borrow, to buy, to marry for, to settle upon,
to commune with an entity which, in truth,
cannot possibly be owned, by anyone, ever.

But it remains a reality that, in most societies
today, ownership of land happens—even if the very
concept of ownership reduces itself to one simple and
popularly accepted fact, as a land lawyer put it: own-

ership means *that you have the right to call the police to throw anyone else off what the title documents say belongs to you. It is not so much that you own it; it is just you own the right to tell everyone else to keep off it.*

There is rather more to it than that, naturally. There is the so-called Bundle of Rights, an aspect of land-ownership recognized in law at least by most western societies. You have the right of possession; the right of control; the above-mentioned right of exclusion; the right of enjoyment; the right of disposition. All of these are legal terms that extend in law far beyond their common-sense dictionary definitions, and which, to the enduring pleasure of the legal community, are still open to all manner of shades of interpretation. And 'twas ever thus.

To begin the process of possession, control, exclusion, enjoyment, and disposition, once the very idea of ownership is accepted, as it has to be in the pages that follow, then a single cardinal rule applies. In order to own the land, you need to know where it is. You must know where it begins and ends, where its boundaries are. The land needs to be delineated and demarcated, and with great and unerring precision. Whether it is a meadow or a nation, a pasture or a principality, its size and shape and the length and direction of all of its edges need to be known.

Enter, some thousands of years ago, men equipped with ideas, devices, and machines specifically designed to do just that.

PART I

Borderlines

1

When the Worm
Forgave the Plough

Nomadic shepherds, accustomed to the age-old
traditions of the Island, would find the ploughman,
his queer implement and his docile beast, profoundly
disturbing. But these first ploughmen . . . had come
to stay and to prosper, and to sow in strange straight
furrows the seeds of a new civilization.

—DONALD R. DENMAN,
Origins of Ownership (1958)

Those who today make or collect or are enthralled
by maps like to say that cartography is a very
much older calling than agriculture. It is an assertion
hugely difficult to disprove: no maps of such antiquity

have ever been known to exist, even scratched on bone, or walls, or traced in preserved mud. Logic of course suggests that primitive cartography was an essential. An early hunter would have come to accept the facts of night and day, of sun and moon and stars, of the coming and passing of the seasons. And he also had to come to know, if only for his own survival, some details of his orientation—of where he was, in relation to everywhere else around him. Otherwise if he hunted, how would he find his way home? If only in his mind, he had to have a map.

In good time such a map would indeed be drawn, and once it was—even if then still in the mind and its lines not yet traced onto parchment or papyrus or scratched into sun-dried mud—it had in good time to include one purely human construct, a construct that would sooner or later lead to the idea that some early peoples would come to forge a relationship, in one form or another, to the land from which they drew their sustenance. To the idea that some people, sooner or later, would come to possess the land, to own it.

To realize this construct, consider two farmers, back in the day. Consider in particular a pair who were living close by each other in the southern part of what is now England, four thousand years ago, during what most regions of the world think of as the late Bronze

Age. The time well after the Stone Age, the era before the discovery of the usefulness of iron.

These are settled people, not nomads. They may best be described as arable farmers, typical of many others around the world who at the time were working their particular pieces of land, the greater number of them growing cereals: wheat, corn, barley, millet, einkorn, bulgur, alfalfa, cotton, rice—the particular grain they sow and tend and harvest depending on where they live and work and have their being.

Since these men are in England, they are probably growing an early strain of wheat: seeds found in a cliff on an island off the south coast have recently yielded DNA samples suggesting so. Maybe elsewhere, and at the same time, similar farmers would be laboring beneath the date palms of Nile Valley, or broadcasting their seeds in the Mesopotamian wetlands between the Tigris and Euphrates Rivers. Maybe they were scattering grains in the Great Rift Valley of eastern Africa, somewhere between Olduvai Gorge and the continent's second river, the Limpopo. Or perhaps they were planting in the loess barrens of early China's Yellow River, the Huang He.

Most probably the two Englishmen, if such they can be called, are known to each other, are friends. Assuredly they are neighbors, and can see one another as they work. Perhaps they are occasional rivals, mostly at harvesttime—although the concept of harvest and harvesttime was relatively new, four thousand years ago. Most societies before that time enjoyed an existence that was largely based on chasing and hunting animals and on gathering wild plants. Humans had not yet fully experimented with the idea that a man might dig into the land and in the exposed soil might plant seeds himself, and that with the passage of time and the variations of temperature and lightfall and the coming of rains might notice how the seeds he had sown changed and grew and eventually produced fruits or grains or edible leaves.

Similarly with animals: it would have taken some time before a settler might realize how best to capture and keep rather than catch and kill. He also might notice that in time, with caught animals, he might even be able to see the mechanics of breeding and the nurturing of young. He might then bend these creatures somehow to his will in a variety of ways, for giving meat or milk or becoming beasts of burden or being employed in draft.

In sum, by around four thousand years ago man would in most parts of the world have abandoned the dangers of following wild creatures across a landscape in the hope of slaughter, and of picking edible vegetation while passing through. The concept of animal husbandry was being born; the concept of settled agriculture was begun. And with this latter, so came the start of the demarcation of land. The establishment of its boundaries. And the realization of the importance of knowing how one man's land is made identifiably separate from that of another.

The use to which a piece of land is to be put is of no consequence. It really does not matter whether the land is to be tilled, as here with these farmers, or grazed or cropped or quarried, whether it is to be built on or mined or left to lie fallow. All that is essential is that it has to be defined. Its whereabouts have to be known. Its dimensions have to be agreed. Especially its edges, where one piece of surface abuts onto another.

In other words, the piece of land has to have boundaries. It needs agreed-upon borders. It must know its place. Place may of course be many things—it may be sea or earth or a spot in the firmament or on some distant world. And place may in many cases actually be land. But land is in all cases, and *always*, place. Before anyone like these two farmers can lay any claim to it, so this place must be somehow distinguished from all other places. It has to be unique. It has to have position. No two tracts can occupy the same position, no two or more can be in the same place at the same time.

Schrödinger might argue otherwise, but the land is not a cat, cannot truly be there and not there at the same moment. Land's place is essentially fixed, and that fixity needs to be established and be properly and literally *demarcated*. It would have been these first farmers who worked out the best means of doing so.

Their ancestors—early Bronze Age men, maybe even those living in the late Stone Age, the Neolithic—would have been among the first to have broken the soil for planting. They would have done so with the use of fire-hardened digging sticks—caschroms, as students of prehistorical agriculture know them today. First a settler—settled now, no longer wishing the nomad life—would have made a single small indentation into the earth. Experimentally, wondering what might happen, he would have scattered a handful of seeds into the cavity, then smoothed the soil back over them. After this, and with time and patience and rain and sun, so came the glorious epiphany of his seeing a green shoot, a seedling, some magical entity that sprouted above the patted surface of the earth. Plant life created at his own behest! Botany, no longer merely accidental!

Later on he would extend that first small indentation—he would use his caschrom to indent the soil once more, but this time in an elongated style, which with effort he could make into what we now know as a

furrow. More scatterings, more patience, more rain—
and that first single seedling would in time become a
straight row of plants, the bounty of just the one seed
presently becoming a harvest of many. He would be
able to collect the grain or the leaves or whatever part
he wished, and with the bounty feed himself and those
around him, maybe his animals, too. And thus was born
in this one place a system of settled agriculture, and once
begun it would start to accelerate rapidly.

Improvements came fast. The design of the caschrom would slowly evolve—one stick would become
three lashed together with a vine, then pieces of harder
and sharper materials were attached to help break more
deeply into the soil—and then as the Bronze Age passed
into the Iron Age, so such implements as had business
with the earth itself were no longer just whittled sticks,
but became tools hammered and forged from smelted
metal, were made stronger and sharper and more durable. This time saw the birth of the plough and the
ploughshare—among the key early components of civilization, lately regarded as standing alongside such other
later seminal inventions as the stirrup, the compass,
gunpowder, and, in due course, the printing press.

The more sophisticated matters of traveling and exploring and blowing things up and disseminating news
could be left for later, though. For now the simpler

A crofter making a furrow in his Scottish smallholding using a caschrom, a sharpened-stick precursor of the plough.

business related entirely to the land, then even more than today an essential and elemental feature of human existence. With some of the plough's critical parts, the business ends, by then made of metal, so the friable and fertile earth of which the land here was made could and would be incised with deeper and straighter lines. Soon the crucial invention of the plough's *moldboard* would arrive, and which would allow the soil to be turned as the plough moved through, with the turned earth duly aerated and made more productive, the resulting crops richer and taller and stronger and fuller of yield.

One further development remained—the harnessing of the captured and husbanded animals to these newly invented ploughs. Once the necessarily complex binding arrangements of leather strips and rope had been fashioned, so any number of tamed and docile but powerful four-legged creatures—oxen, mules, horses, donkeys, in some far places camels, all of them stronger and more enduring than humans—could be bent to human will in this new and previously unimagined way. The beasts could be persuaded to pull the ploughs behind them and allow their abundance of muscle to help with the digging of the furrows, and then for the planting of seeds and the harvesting of crops to be accomplished ever more extensively and effectively.

In the meadowlands of ancient southern England, the period saw the assembling of what some anthropologists have called the Deverel-Rimbury plough teams. Here in England, as in the Fertile Crescent and the Great Rift and most especially in China, where the sophistication of early agriculture predated and then outstripped all others, the business of field systems and field design begins and, with such thinking, so the true start of land demarcation.

Here, boundary lines begin. Our two farmers—for the sake of clarity, let us say a pair of these Deverel-Rimbury age farmers, of a people who were otherwise best known for their distinctive globular pottery and their burial barrows. These men tilled their lands close to a settlement in a deep valley among the rolling chalk and limestone uplands of what is now Dorset and Wiltshire. They are occupying adjoining swathes of this land. Let us imagine that one of the pair is planting his crops on river bottomland, while his colleague has located himself slightly higher on a hillside. Maybe because of the rather different terrains—or maybe because of personal whim—the two farmers decided to create furrows that run in slightly different directions.

Let us imagine as well, for sake of argument, that the man working the bottomlands arranged his furrows to run in straight lines to the north and south. Let us also suppose that his neighbor up on the hillside ran his furrows along the contour lines, that he created small terraces to keep the strips between each terrace more or less flat, so that he was not planting on earth that lay at an inconvenient angle. The relics of such strip lynchets, as these are called, are to be found in meadows across southern England today, almost hidden memorials to the ancient field systems of four thousand years ago, and fascinating and illuminating in themselves.

But in terms of demarcation, the intricacies of lynchet and furrow design are not so important. What is crucial for the origination of boundaries is the simple fact that where in this case bottomland meets hillside, where the newly ploughed furrows of the first farmer meet the strip-lynchet lines of the second, there is an obvious disconnect—what in geology would be called an *unconformity*. Furrow lines of the first farmer, all proceeding in a north–south direction, suddenly encounter furrow lines of the second farmer, which run in an altogether different direction—perhaps east–west, perhaps at a diagonal.

And this point of unconformity—which may at first be marked by a line of pebbles or stones or boulders, or by planted standing sticks, or even in due course by a crude barrier fence of wattle or a hedge of blackthorn—

Evidence of ancient man-made earth terraces known as "strip lynchets" are still to be found in southwest England.

turns out to be a boundary. The first-ever mutually acknowledged and accepted border between two pieces of land, pieces farmed or maintained or presided over—or owned—by two different people, in this case neighbors and friends.

By the simple act of each farmer creating furrows of his own peculiar alignment—alignments either forced upon each man as a consequence of topography, or simply engineered by personal choice—so the land, for the first time, had been informally demarcated. Later on, the informal would become the settled, and then formalized. A frontier had been established. The basis of land possession had been established, and the whole complicated engine work of landownership now had been unwittingly handed its origin myth.

The two Bronze Age farmers gave birth to a system that would lead in time to the making of borders not just between individual people, but between villages, between towns, between counties and prefectures, between states, between nations. The formal fault lines of the world have their origins in the informal decisions taken by a pair of prehistoric seedsmen in Wiltshire, and by other plough teams in the Yellow River Valley, in Mesopotamia, in the pueblos of New Mexico, in the mudflats of Bengal, and in the paddies of central Japan.

The surface of the world is divided in numberless ways, and at a variety of scales. At the most granular and personal level is the division like this, at the level of mere fields. Such lowest-level boundaries inform the personal lives of millions, so long as they are landed to even a small degree. You are aware of such boundaries if you are a householder in a Dorset village or in a suburb of Denver or, nowadays, in Wuhan or Sarajevo or Manaus. A Bantu cowherd knows where he may graze his cattle in a kraal, and where he may not. A rice farmer in Gifu Prefecture in western Japan will show the utmost respect to the position of his neighbor's paddy, and not stray beyond uninvited. A sheep baron in Queensland or a bison farmer in Wyoming will know the exact extent of his station. The size of each personal possession may be as varied as its function, or as the required bragging rights of its possessor: it may amount to mere fractions of an acre, or it may be measured in tens of square miles of prairie and range and take endless hours for a vehicle to cross. Each holding will however be firmly delimited by the presence of a borderline—most probably unmarked for the larger properties, though with stone walls and then wooden fences and Leylandii hedges and electric wires as the scale diminishes.

Populations of such individuals will also form clusters, will inhabit regions and cities and towns, which are defined by borders too, borders generally drawn to a larger scale and with more weighty implications. Within a garden fence there is just the simple matter of ownership; within an urban boundary there is usually the concept of jurisdiction, there is tax collection, there are powers of police and provision of services, there is the nuanced complexity of governance. To a lesser extent, in villages and parishes, there are metes and bounds that have been established for centuries past— though all evolved from ancient field patterns—and for which there may be traditional reminders, ceremonies and celebrations, held to prod those within the pales of settlement to understand the existence of limits to the territory, and that these limits merit respect and preservation.*

---

* Traditional "beating the bounds" ceremonies still take place in many English villages and towns—and in some few North American settlements too—each springtime, usually around Ascension Day, designed forcefully to remind the citizenry of the location of their boundaries. The rituals invariably involve what is said to be a necessary degree of minor cruelty: with the vicar leading, prominent townfolk armed with canes will lash their way along the boundaries, and young boys are usually brought along so that at each way station they may either be mildly whipped or have their knuckles rapped to remind them,

And as with concentrations of humanity, so too
with the greater area, the hinterland, the demesne: so
in any country of any size there are counties and prefectures and wapentakes, dioceses, sokes, hundreds
and boroughs. To measure and map and demarcate all
of these—which may well already have become self-
delineated by virtue of differences of language and
law, dialect and customs, even by variations of mores
and philosophies—required the labors of teams with
tripods and theodolites, to set and codify and make
permanent boundaries that would likely last for generations, even for centuries.

Wider still is the matter of the land that encloses or
is enclosed by whole nations. The question of how and
why nations and nation-states come into and pass out
of existence belongs elsewhere—but the matter of the
delineation of these entities, often very large entities
containing multitudes, both of peoples and of acreages,
is part of this account. There is the enormous business
of official national surveys, and then still beyond there
is the matter of the making of internationally agreed
boundaries. Such invariably involves a pair of parties,
two separate aggregations of peoples who, though they

---

for future generations, where the borderlines run, where they
turn and where they end.

The traditional ceremony of beating the bounds is still
performed each springtime in many British communities—
a young boy often forcibly reminded of the site of the local
boundaries.

are technically neighbors, may possibly be of different
natures and ethnicities and languages and politics, and
who may disagree with one another and may do so with
sufficient vehemence occasionally to lose their mutual
temper and fight—these differences present special
challenges to those who would define and mark out the
borders between them.

And more challenging yet is the matter of the defining and marking of frontiers, which are necessarily still

morc grand and yet often more vague—boundaries between entire masses of national entities and ideologies and religions and outlooks. The Iron Curtain was one such frontier, though it has generally evaporated now. The present division of the Middle East that now divides Israel from her frontline neighbors can fairly be defined as a frontier rather than an uncomplicated border. And then again, there is the visible-from-space-at-night line of lights and wire that keeps India from Pakistan, and with just a single, spectacularly theatrical border crossing in the Punjab between Amritsar and Lahore—that too can rightly be called a frontier.*

Beyond the frontier still, but still philosophically related to those farmers back in the Bronze Age, there is the matter of the world at large. The very notion of the ownership of everywhere depends ultimately upon the precise placement of every single one of the world's

----

* The idea of the frontier here was long memorialized on the permanent way by the Frontier Mail, one of the great Indian express trains. Originally run by the Bombay, Baroda and Central India Railway, it was a daily service between Bombay Ballard Pier—connecting with the P&O steamers—and Peshawar, just below the climb to the Khyber Pass (beyond there was a smaller, steam-hauled railway to Landi Kotal, at the Afghan border). Peshawar is now in Pakistan; there is no direct train connection with India. And there are no passenger trains at all in Afghanistan.

acres within the 37 billion of the total terrestrial landscape. It just requires science to step in, to work out the immense size and scale of the planetary sphere itself. The task is of course prodigious. To pace out a suburban garden is one thing, to measure a country tract another. To parse the entire planet is to enter quite another sphere of computation.

Except that nowadays, with measurement so very precise, all measurement of all scales are related and relatable. To be able to know and measure and chart with the highest degree of accuracy the tiniest details of the world—the size of the village green in Wiltshire, the location of the oasis outside Mashhad, the slope of the old volcano in Rotorua—one also has to have an absolutely accurate notion of the size and shape of the planet itself. It was a Russian tsar, Alexander I, who in the early nineteenth century spearheaded and financed the first modern effort to try to do such a thing. He paid sizable sums in rubles to send out into the field parties of skilled surveyors, charging them with one outwardly simple task, which continues to be posed today. They were to try to determine just how big is our world: What exactly is its shape? How does the land lie, where does all the land exist?

2

The Size of All the Earth

Unbounded freedom ruled the wandering scene
Nor fence of ownership crept in between
To hide the prospect of the following eye
Its only bondage was the circling sky

—JOHN CLARE,
"The Mores" (1837)

Finding one of the survey marker points for what is now called the Struve Geodetic Arc in central Latvia turned out to be a little less easy than I imagined. The rented car was old, its driver a little unfamiliar with the territory, it was raining, and we managed to get a flat tire. But by great good fortune there was a farm nearby and a couple of Britons who raised alpacas on it, and they had the necessary tools and a jack. After we had bolted on the spare wheel and been fed on strawberries and tea—and had petted the alpacas, perhaps not an everyday occurrence in contemporary Latvia—the rain eased somewhat, and then we were off again into the woods. Eventually we came across a green metal notice by the dirt road on which we were traveling, and behind it was a footpath that led up a hill to a clearing.

The sign identified this otherwise sparsely inhabited and unremarkable spot in the woods as lying in the parish of Sausneja, on the far outskirts of the town of Ergli, in that middle part of Latvia once known as Livonia. This was a part of the Baltic countryside that had been battled over for centuries by, in particular, the armies of Germany, Poland, Russia, and Sweden. Locally all had been peaceful for some recent years—although painful memories of the Nazi occupation and the fifty years of Soviet rule that followed were still fresh.

There is constant local fretting today that the Russians might well soon return, laying claim to lands that some in Moscow and St. Petersburg have long believed still rightly belong as a crucial Baltic component of their empire. The average Latvian is thus only too well aware that the farmland that makes up most of his country has been all too frequently demanded by others, has been fought over and requisitioned by outsiders, and is still claimed by foreign rivals who see the dwindling presence of the two million Latvians—whose distinctive language, culture, and music have survived centuries of tussling and threatened eradication—as a mere inconvenience. The historical fate of land worldwide could easily be illustrated by the situation of Latvia's sixteen million acres: they have been variously bought, sold, owned, seized, fought over, confiscated, and ultimately parceled out to private owners, a melancholy parable of territorial ambitions and desires. And yet right here, in the middle of this Baltic nowhere, is something belonging to all the world.

On the footpath leading up into the woods the grass was slippery from the showers, and clouds of mosquitoes frothed languidly under the dripping branches. After a quarter of a mile, at the low summit from which through the thinning trees we had a fine view of the meadows, small lakes, and rivers below, there was what I had hoped for: a three-foot obelisk standing guard beside a flattish alien stone that was half-buried in the ground, almost invisible among the grasses. The stone—granodiorite by the look of it and probably brought in from the hills up near the Estonian border—had a large X crudely incised on its weathered surface. It was a marker, said a plaque on the obelisk, that had been placed there in 1821. It was one of two surviving relics in Latvia of what since 2005 has been designated a United Nations World Heritage Site, as illustrating one of the first and most important attempts to establish with great accuracy the size and shape of Planet Earth.

It was the Greek librarian-scholar-astronomer Eratosthenes who famously first computed the circumference of our planet, some 2200 years ago. He did so by comparing the angle at which the sun's noontime rays fell onto the water wells in Alexandria, his Egyptian hometown, on the same day that they fell vertically on, and thus illuminated the very bottom of similar wells in Aswan, some 524 miles upriver along the Nile. The shafts of sunlight in Alexandria were slightly off vertical, and Eratosthenes suspected that this was entirely due to the curvature of the Earth, if indeed the Earth was the enormous ball-shaped planet that many since Pythagoras, Plato, and Aristotle had suspected it to be. He measured the angular difference at 7 degrees—about one-fiftieth of the 360 degrees of the sphere. It followed, he reasoned, that if 524 miles was a fiftieth of the total, then the circumference of the sphere would be some 24,000 miles, nearly 39,000 kilometers—not too far from the roughly 40,007 kilometers that satellites, lasers, and GPS devices declare it to be today. For this hugely significant and pioneering realization, Eratosthenes of Cyrene, equipped only with a fine and logical mind, a protractor, a set square, and a plumb line, now occupies a deservedly permanent place in history.

But fast-forward two thousand years, and to the more rigidly scientific minds of the early nineteenth century. A number of scientists of the time—notably many of them French, with names like Picard, Bouguer, La Condamine, Delambre—grappled with the knotty problem of ascertaining the Earth's size and shape: Was it a sphere or a spheroid, oblate or prolate? Was it flattened at the poles, broadened at the equator? Their attempts mainly involved as precise as possible measurements of short stretches of the planet's meridians, a few miles long—and the results were good, but not convincingly so. There had been some moderate success, for example, with surveys conducted in the late eighteenth century in Peru and Lapland, but the equipment available was comparatively crude, the lengths of the meridians unsatisfactorily short, the results unconvincing.

For the results to be entirely satisfactory to the geodesists' community it would require the measurement of one very long meridian line, hundreds of miles worth, and with exceptionally accurate instruments to make the measurements and perform the calculations. It took the chutzpah of a Russian, and the foresight of two imperial tsars, ruling at the very height of their empire's global standing, to confront the problem head-on, to disdain the work of the ancient Greeks, and to dismiss what those in the Russian capital regarded as the halfhearted shillyshallying of the French measuring platoons and of those who had similarly tried in South America and the Scandinavian Arctic.

One of the preeminent astronomers of the time, living in what was then the Russian possession of Estonia, was Friedrich Wilhelm Georg von Struve, of a family who—rivaled, perhaps, only by the Herschels*—uncannily produced five generations of highly competent astronomers. It was Struve who decided that, for

---

* In the mid-eighteenth century the German-born but London-based astronomer William Herschel and his sister Caroline between them discovered a score of comets, twenty-five hundred nebulae, and, most notably, the planet Uranus. William's son John became in turn so accomplished in the field that he was buried in Westminster Abbey beside Sir Isaac Newton. And Alexander Herschel, John's second son, was elected a Fellow of the Royal Society and was the world's leading nineteenth century expert on meteorites.

the glory of his empire, it would fall to him to measure the world properly, and to do so with the greatest of precision. And since to do so required the measurement of the exact length of one of the planet's meridians, or lines of longitude, he would do so along a track conveniently close to where he lived in Estonia, some 25 degrees east of Greenwich. (It is probably needless to say that only the measurement of longitude-line meridians will give the size of the world, since all those meridians pass entirely around the world. By contrast only one line of latitude—the equator—girdles the entire planet; the others—such as the Tropics of Cancer and Capri-

A series of small markers, running in a line from Norway to the Black Sea, show where Friedrich von Struve measured the meridian to determine the exact size of the Earth.

corn, or the Arctic Circle, are very much shorter, and encompass only fractions of the planetary sphere.)

The task took Struve fully forty years. The Struve Geodetic Arc, by which he is now internationally memorialized, is one of the greatest—but also one of the least remembered—of the scientific achievements of the age. The traces of its progress are to be found passing through ten countries now, from the near Arctic north in Norway down to the warmth of the Black Sea west of Odessa in Ukraine. And passing, as it happens, through the alpaca-rich Latvian hamlet where the X is marked on an almost hidden piece of stone.

Struve was a young and exceptionally energetic man* when he formally began the survey at his observatory in the Imperial University in what was then the city of Dorpat—now Tartu—in Russian Estonia. His idea was to chart the exact length of his meridian running for about 1600 miles—and from this both derive the size and shape of the planet and by doing so enable the making of ever more accurate maps—a growing necessity since the private ownership and distribution of land was becoming ever more commonplace.

He decided he would begin his survey beside the marble front doorstep of the yellow-and-white painted

---

*   He fathered eighteen children by his two wives.

Dorpat Great Observatory, of which he was director—
and which, thanks to the generosity and scientific curiosity of the Tsar Alexander I, he was busily stocking
with a formidable and most necessary collection of
telescopes and surveying equipment.

One summer's day in 1816, Struve and his fellow astronomers hauled out of the observatory basement the
key elements of this stupendous accumulated arsenal
of measuring devices—in prime position an enormous
German-made theodolite (one of Struve's surveyors
had seven theodolites and would choose the one that
best suited the reading he wanted to take), a twelve-
foot-long telescope known as a zenith sector, a giant
brass quadrant, and a set of precisely made surveyor's
chains. With infinite care and exactness and working with a team of Dorpat students, Struve used these
and a variety of tripods and drills and tower-building
equipment to measure first a long baseline between the
observatory and a local hill, and from which he would
then begin his historic triangulation.

As its name implies, triangulation has everything
to do, and with elegant simplicity, with the realities of
the geometry of triangles. With the idea that if you can
know the exact length of one of a triangle's sides, and
carefully measure the three angles of the figure, then
the lengths of the other two sides can be calculated

with dispatch. This unadorned fact of plain geometry has enabled the measuring, planning, and making of just about everything concerned with large tracts of raw and unimproved land—from the building of the Pyramids of Giza to the accurate placement of the Iron Curtain—and to the measuring of the size of the world.

So, once Herr Struve had carved his letter X into his observatory doorstep, he performed two specific acts. First—by employing a sextant and a chronometer—he determined the latitude and longitude of his initial position. Next he created a baseline that originated from the center point of the X. By using either finely calibrated wooden or metal bars, or more usually special chains—Gunter's chains most probably, each one 66 feet long and with 100 links, made of a brass that was little affected by changing temperature and would neither stretch nor shrink—he then determined the length of this baseline, with accuracy to the smallest possible fraction of an inch.

He would next erect his Munich-made Reichenbach theodolite over one end of this line—with the instrument's tripod center point poised exactly over the baseline's measured end point—and using the graduated graticule of his theodolite telescope he would peer out across the landscape and select a prominent distant point, a hill or a steeple or a tree—and mark a notional second line all the way to that. He would then lug his theodolite and quadrant and zenith sector—or have his staff help him do so, since these instruments were formidably unwieldy and heavy things, enormous confections of brass and glass and wheels and gears and all mounted on metal-bound ash tripods—to the other end of the baseline: the doorstep end, where the triangle was begun. And from the new mounting point on top of the baseline's exact origination point, just above the scribed letter X, he would peer through the lens and draw a third line to the self-same point (the hill, steeple, or tree), which he had seen from the line's other end.

By these somewhat cumbersome means he would have plotted himself a triangle. By doing this, he would also have drawn himself three angles—all of which would, by the ineluctable magic of Euclidean truth—add up to 180 degrees. By precisely measuring the three angles, once again with his theodolite, Struve would then be able to calculate the exact length of the other two sides of the triangle. They would be as accurately measured as the baseline—but this time not with chains, but with calculated mathematical deduction.

This, then, is the basic magic of triangulation: that from measuring the exact length of just one line and then afterward drawing two more and measuring the resulting trinity of angles, the length of the two other lines—which would hitherto be unmeasured distances between places—could be worked out. Geometry was all: the notion of having to chain every line—through jungle, across mountains, or up slippery wet grass slopes in Latvia—was now entirely unnecessary. Simple mathematics—and formidably precise instruments fashioned with finely chased brass, lubricated bearings, well anchored and highly stable footings, and impeccably polished and perfectly shaped lenses were key—produced elegant and accurate answers, time and time again.

It was slow and ponderous work. In total, the surveyors in these hinterlands of the Russian empire found it necessary to construct 258 main triangles, and for checking purposes to measure ten baselines—using varieties of chains and metal and wooden bars that, it was claimed at the time, could give an accuracy of length as great as one part per million. The men had to haul their immense instruments over swamps and lakes and rushing rivers, through forests and snowstorms and, in the wilds of the Scandinavian north, across brutally gale-swept ice fields.

To make one set of triangles across the Gulf of Finland the men had to clamber up the inside the tiny spire of an ancient church, then build a platform among the bat-infested rafters, aim their brassbound telescope across the body of water, and peer through the sea fret for the pinpoint of flashing light from a heliograph that had been set up on a tower they had specially erected on the far side. Tower building was indeed a major part of the survey work: so often there was no available hill or steeple or lighthouse, and so a rickety wooden tower had to be constructed and the equipment hauled up it with ropes and systems of pulleys—the construction having the added benefit of allowing the triangulators to see above the tree line, for the spruce and Scotch pine of the northern forests tended to block the view.

The project proceeded in fits and starts, its progress depending largely on the generosity of the St. Petersburg treasury on Nevsky Prospekt. It received a great fillip in the late 1830s, when Nicholas I, having assumed the role of tsar following the death of his brother Alexander, felt sufficiently in control of his empire* to demand the building of a new observatory in Pulkovo, just south of the Winter Palace. Tsar Nicholas, who

---

* The fact that a secretive revolutionary organization tried to topple him from the throne as soon as he assumed office in December 1925—the group thereafter being known as the Decembrists—left him understandably nervous during the first years of his reign.

never anticipated he would rise to the throne, was an engineer by trade, was fascinated by machinery of all kinds. Struve, now famous and honored by astronomical societies around the world, would be invited over from Dorpat to be the director of the new Pulkovo Observatory; he would be allowed to buy the best and the biggest astronomical devices then known—including a then legendary 30-inch refracting telescope built for him by the firm of Alvan Clark and Sons, close by Harvard College in Cambridge, Massachusetts.

And sufficient money would now be pumped into the meridian project to complete it. As it duly was, in the summer of 1855. The line was at last fully measured, Hammerfest to the Black Sea; and from its length could be deduced the length of a quarter meridian, pole to equator. It had never before been done, and it produced metrics of an accuracy never before known.

And thus the planet, it was calculated, now had a Great Circle of a carefully derived size and shape. Each quarter meridian, Struve's teams declared in a massive tract that he published that summer, was worked out as being exactly 10,002,174 meters long. Its circumference was thus four times that, or 40,008,696 meters.

It was staggeringly accurate. By comparison: the latest figure put out by NASA, based on satellite measurements, shows the planet coming in at 40,007,017 meters round. Eratosthenes, two thousand years before, had calculated the circumference as being, in today's units, 38,624,000 meters. Herr Struve's globe was larger than both, but not by much.

With the world's size now so carefully worked out—and a clutch of the world's senior surveyors met in Paris in 1883, to begin the measuring of another nearby meridian that would head even farther south, all the way from Cairo, through the jungles of east Africa to the Cape of Good Hope, to make the calculations more perfect still—the team left for other work. The physical evidence of their art would remain, however. The actual reference points of the scores of huge European triangles, so precisely made—with long holes drilled into rocks, molten lead poured inside, steel bolts fixed into place as the lead cooled, brass plates secured on top, all to produce impeccably accurate foundations for the instrument—were left to suffer in the cold and heat of Finnish winters and Ukrainian summers. Weather, vandals, hunters (who liked to steal the lead to make into buckshot), and fast-growing vegetation gnawed away at the relics of the arc's completeness. By the time, in 2005, when the United Nations decided that what remained of the Struve Arc be made a World Heritage Site—alongside the Grand Canyon, the Pyramids, Westminster Abbey, and the Sydney

Opera House—only thirty-four of the original points could be found. The U.N. officials duly saw to it that monuments to Struve should be built all along the line, with big marble memorial columns at the two termini, in Hammerfest up in the Arctic north and the village of Staro-Nekrasivka down on the Black Sea coast. Ten countries—Norway, Sweden, Finland, Russia, Estonia, Latvia, Lithuania, Belarus, Moldova, and Ukraine—now play host to Herr Struve's monumental achievement.

But few tourists ever come to see the markers. The British alpaca farmers who live by the little town of Ergli in central Latvia knew well of the arc's existence, but thought that the last time anyone had clambered up the rain-slick grass to see the tiny stone with its letter X carved into it back in 1821 had been six months before. It seems that despite the best efforts of a few, Herr Struve is a hero still somewhat overlooked, a prophet without honor, even close by his Baltic home.

3
Just Where Is Everything

The world is a great wilderness where mankind have wandered and jostled each other about from the creation. Some have removed by necessity and others by choice. One nation has been fond of seizing what another was tired of possessing; and it will be difficult to point out the country which is to this day in the hands of its first inhabitants.

—HENRY ST. JOHN, VISCOUNT BOLINGBROKE, *Reflections upon Exile* (1716)

Now, with the size of the world fully determined, it remained only to make a map of it. A map of all the world, as precise and as a perfect as may be.

This too was as almighty a project as it sounds, conceived at a gathering of geographers in Switzerland in 1891, just as the African arc geodesists had assembled up the road in Paris eight years before and as their Russian predecessors had gathered in Dorpat earlier in the century. This new conference was held at a time when geography and cartographic technique had evolved into a very much more advanced science, and for which the girdling of the earth was a quite vital matter. It was also a time when European empires—particularly the British, the French, and the German—were at their apogees, and yet when most physical geographers, even when from one of the imperial powers, maintained a studiedly disinterested view of their craft. Physical geographers back then, as mostly still today, took pride in remaining as politically neutral as the land was itself, caring little for which nation ruled what, only for the nature of the world's fantastically varied surfaces and their physical appearance and settings, the location of the planet's river and peaks, its deserts and swamps.

Albrecht Penck was a scholar preeminent in the field, an expert on glaciers and the Alps and the geology of the Pleistocene. He was a German, though at the time of the intellectual epiphany that gave birth to the world map project, he was settled in Austria, a professor at the University of Vienna. It was an appropriate city from which to witness the birth of so noble and yet ultimately so ill starred an idea: in the 1890s the Hapsburgs and their kin, rulers of their immense Austro-Hungarian empire, were blessed with a peculiarly cosmopolitan view, endlessly curious, intrigued by everything in the world around them. The Hapsburg archduke Franz Ferdinand, still twenty years away from meeting his world-shattering assassinated end in Sarajevo, was about to set off on a celebrated world tour, which saw him hunting kangaroos in Australia and taking himself, locally unannounced, to such unfamiliar places as Sarawak and Hong Kong, Bangkok and Nagasaki. The Viennese, sedate in their coffeehouses and comfortably taking their morning chocolate at Demel, read avidly about every encounter between the incognito archduke and his hardly incognito retinue of chamberlains and hussars, on the one hand, and on the other the bemused locals of his dozen or so Pacific stopovers.

When Franz Ferdinand had initially set off for Trieste and the twin-funneled cruiser that would transport him and his suite to the Pacific and then around the world, the Hofburg announced that his adventure was specifically designed for his "information and study."

Albrecht Penck, the cartographic visionary who very nearly succeeded in mapping the entire planet at the scale of one to one million.

The Austrians, following his travels reverentially via the lengthy dispatches in their morning newspapers, found themselves more than happy being educated along with him, gleefully coming to know more about the outside world than did most in other landlocked nations, especially considering that this one had only a rudimentary navy and a merchant marine that was wanting in sailors who might bring home stories of the exotic outside.*

---

* The international press of the day still had an appetite for the exotic, with Sydney's *Town and Country Journal*, in covering the archducal visit, also telling its readers the stories of a boxing

It was armed with this kind of peculiarly Austrian geographical enthusiasm that Professor Penck traveled through the northern Alps to Berne in 1891 and presented the conferees at the Fifth International Geographical Congress with what seemed at first a wildly romantic and impractical dream.

He declared that he wanted the delegates formally to agree that the entire world should promptly be mapped, with everywhere on the globe—every continent and every nation and every island owned or unincorporated—drawn on permanent paper at the scale of one to one million (a little under sixteen inches to the mile), and that all the resulting maps should be of the same style—using the same systems of colors, the same units of measurement (meters), the same spacing of contours and hypsometric tints (darker shades at the bases of hills and mountain ranges, paler as the hills became higher, white where there was snow), the same system of typefaces (employing the Roman alphabet—no Cyrillic, no Arabic, no Chinese), and the same language (English).

---

kangaroo seen in London and, across in Oxford, of a six-year-old black-maned Caffrarian lion who was brought into a theater in a long cage and then released to wrestle with his trainer, "a stalwart negro."

Moreover, the maps' sheets should all be limited to a rectangle of four degrees of latitude and six of longitude, and their projection should be a kind that if all the sheets were to be fastened together (clear tape not at the time invented) they would form themselves into a gigantic sphere exactly one millionth the size of Planet Earth. An oblate spheroid of globe, in other words, as Herr Struve had lately proven, and which at Penck's proposed universal scale would be about the size of a moderately large house. He would call the project the International Map of the World, the IMW—*a common map*, as he put it, *for a common humanity*.

It was something of an iconoclastic idea for the fractured and fractious world of the nineteenth century, and the geographers present that day applauded with enthusiastic surprise. Hitherto all too many maps of all too many countries had been jealously created with a mind-set of defensive territoriality, the mapmakers shielding their work like adolescents in an examination room, cupped hands covering their work from prying eyes. The British were among the worst offenders, promoting the size and importance of their own imperial holdings and displaying a haughty cartographic indifference to lands not under their control. Earlier empires, from the Greeks to the Venetians and the early Ottomans, those that flourished before the technologies of theodolite-based mapmaking had arrived at the end of the eighteenth century, were perhaps not so blameworthy, resorting to making maps that showed blank spaces, or mysterious areas marked *Terra Incognita*, or *Terra Nullius* or *Here Be Dragons* and which displayed more ignorance than indifference. But at the height of British imperial power, with which the Berne conference coincided, a peculiarly English combination of insouciance and arrogance helped to produce a distorted view of the mapped world—a view that Penck was determined to change.

Nevertheless he did install a Briton on the IMW committee—a Scots-born journalist, teacher, and impassioned geographer named Sir John Scott-Keltie, whose eagerness for all to know geography as a science—he helped expand geography teaching in schools, helped establish readerships at both Oxford and Cambridge, urged the Royal Geographic Society to give as much weight to promoting geography as it did in sending expeditions into the most exotic faraway— was matchless, and earned him a chest groaning with gold medals and a knighthood from the King. He was also the longest-serving editor—for forty-four years, until his death in 1927—of *The Statesman's Yearbook*, a massive and legendary compendium of global information still published each year in London, as it has been without interruption since 1864.*

This modest Scotsman was joined on the committee by an American of even greater public reputation: John Wesley Powell, revered father of the exploration of the American West and the first white man to take a boat down the entire length of the Grand Canyon. As with so many reputations, his has lately suffered somewhat: his divisions of the world's peoples into the savage, the barbarous, and the civilized (with the Native Americans he encountered seen as exemplars of barbarity, yet with ambitions to become civilized) has won him few

---

* It was a nineteenth-century British prime minister, Robert Peel, who first urged publication of such a book. Its founding editor, Frederick Martin, set a tone from which it has never varied: "The great aim has been to insure an absolute correctness of the multiplicity of facts and figures given in *The Statesman's Year-Book*. For this purpose, none but official documents have been consulted in the first instance, and only when these failed or were manifestly imperfect, recourse has been had to authoritative books and influential newspapers, magazines and other reliable information." Thus, anyone needing to know the size of the Gabonese navy (seven patrol boats, five hundred sailors) or the acreage of Denmark under wheat (1.8 million) or rye (108,000) need look no further.

twenty-first-century friends. And though he only had one arm—his right was half shot off at the Civil War Battle of Shiloh in Tennessee—and though his rock-climbing abilities were prodigious and his leadership skills in the rapids-dashed Colorado River boats were without parallel, some have claimed his expedition stories were embroidered and bowdlerized. For instance, Powell failed to mention that because of his disability he allowed himself a life belt, while his men, if they went in the water, would have to fend for themselves.

There is no doubt, though, that Powell's view on land use was prescient: he thought that America's intermountain west was quite unsuitable for large-scale agriculture, believing that a chronic want of water was a crucial problem that would affect all of the western states. He also thought state borders—numbers of which were being drawn during his lifetime—should follow natural watershed boundaries, so as to lessen the likelihood of argument over water ownership. He was an impassioned proponent of conservation as well, believing land should in many places remain wild and untouched by humankind, left for the local flora and fauna and the native peoples—even if in his condescending ignorance he saw these as "barbarous"—to enjoy the acreages without disturbance. His standing in the pantheon has yet to fall wholly victim to modern revisionism: there are numberless entities now named for this soldier, explorer, and geologist—there is a Lake Powell, a reservoir, a plateau, a mountain peak, a mineral (a salt of molybdenum, much valued in steel making), several schools, a federal building (the head-quarters of the U.S. Geological Survey, of which he was the second director), a song, a movie, and a small town named Powell which sports six thousand souls in the front range of northern Wyoming.

One would have thought that with such eminent boosters as these the IMW would have gotten off to a flying start. It did no such thing. Division, rancor, petulance (the French displaying much, especially over the use of the Greenwich meridian rather than having Paris as the longitudinal divide between east and west), blind hostility (the British objecting to the use of the French metric system for the maps' scale), and sheer mulishness—all these contributed to a delay of some seventeen years before there was even a signed agreement to begin. And even that was promptly derided as being merely a pious aspiration.

But then serious planning did get under way, if hesitantly. The Greenwich meridian was eventually chosen; the meter was selected; Latin script would be used; a clever numbering system was adopted; and a cunning projection—polyconic, meaning that all of the maps have a trapezoid shape, their upper and lower edges as parallel as the latitudes they represent, their sides (in the Northern Hemisphere) tending to make the map wider toward its lower edge or (in the Southern Hemisphere sheets) toward the upper.

A last wrinkle was occasioned by some spectacular dog-in-the-mangerism—when, for instance, both the Russians and the Japanese declared that each had a prescriptive right to make the IMW maps of China, which was then mired in a postrevolutionary shambles. Not so fast, said China from beneath its wreckage: "we hereby announce that our Geodetic and Topographical Service are now operating regularly in all the provinces of China," and that St. Petersburg and Tokyo might as well mind their own business.

Thus chided, the forty-eight delegates gathered for a photograph on the steps of the Quai d'Orsay in Paris—all long coats and astrakhan collars, impressive facial hair and expressions of high purpose and with the hugely tall Albrecht Penck towering above them, front and center—and told the world they were ready to begin the greatest cartographic endeavor undertaken in all history.

Six months later, Archduke Franz Ferdinand was assassinated in Sarajevo, Europe spiraled rapidly out of control, and by the end of that summer of 1914, the

Great War had begun. Yet as it happened, and some-what counterintuitively, the conflict, at least toward its end, provided a sudden spur to the IMW mapmakers—not least because the reorganization of so much of the world dictated by the Paris Peace Conference of 1919 required maps and atlases to help the world understand where exactly its citizens were now living their postwar lives. Fully one hundred sheets had been produced by the end of the war, most of them by the Royal Geographical Society in London, many of these of British colonial possessions and some scores of the landscape of Imperial India.

During the ensuing two decades the project—which had its headquarters in Southampton, in southern England—swung into impressive action, though with some notable exceptions. France mapped much of fran-cophone Africa. Germany made maps of all German-speaking countries in Europe. The entirety of the former Roman empire was mapped. An agency called the Brazilian Club of Engenharia published a 50-sheet series of Brazil. The Americans began a project of draw-ing 107 sheets of all of Hispanic America, which is still regarded today as "an unsurpassed scientific and artistic achievement." But to general dismay the map-making energy that they displayed toward Cartagena and Mara-caibo was not reflected in their coverage of Colorado and

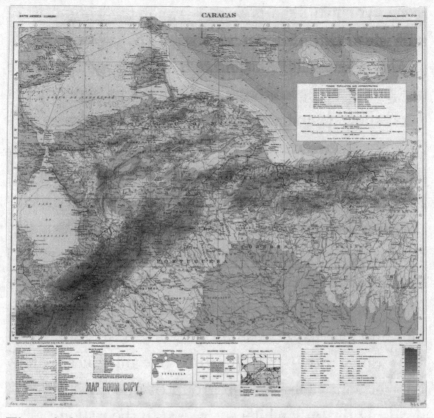

The only known complete collection of the IMW maps is kept at the University of Wisconsin in Milwaukee—almost eight hundred sheets of a project that was tragically never finished.

Mississippi—and Washington's isolationist tendencies halted production of more than a scattering of American millionth-scale sheets. The Russians withdrew from the project after the Bolshevik revolution; and for a variety of other reasons—wars civil and international, institutional incompetence, budgetary problems among them—there were virtually no sheets made between the world wars of northeast Asia, Australia, Canada, or Polynesia.

One signal success stemmed from a joint Chinese and Swedish expedition to map Mongolia and Chinese Turkestan. It spent eight years wandering purposefully through some of the most remote grasslands and deserts of central Asia. This expedition was led by the Swedish explorer Sven Hedin, and with its forty scientists from six countries (mostly Swedes and Germans, since their governments footed the bills) and protected by a force of thirty infantrymen and with its supplies carried across the sands by a team of more than three hundred camels, it was described by some as a "traveling university"—not that this, one imagines, is how villagers en route saw the group of ghostly strangers, probably the only whites ever seen in the region.

The expedition produced four sheets of its proposed trans-Himalayan atlas, with all the information presented in impeccable IMW style—but the huge atlas itself was not to be published until 1966, and, quite bizarrely, because the government of the United States confiscated it. Hedin had wanted it published by a renowned German company based in the city of Gotha— but as fate would have it Gotha fell behind Soviet lines in the partition of postwar Germany. In the days before the closing of that frontier, a team of three officials of the U.S. Defense Department raided the publishing office and hauled off as much material as possible before the Russians could get their hands on it, and flew it back to Washington. The Pentagon decided it could be published only when East Germany's strategic significance was deemed to be minimal—and that turned out to be in 1966, when just about all involved (Sven Hedin included) were dead.

Slowly, slowly, the stack of completed maps increased. The Soviets changed their mind, and as part of Stalin's third Five Year Plan agreed to map the entirety of the USSR—it would take fully 182 sheets—starting in 1940. Stakhanovite labors ensued: the venture was completed—despite the many depredations of war—in 1946, with as a result a further 22 million square kilometers of the world's land surface was added to the accumulating total. The planet's entire land surface area runs to 37 billion acres: the Soviet Union's share was some 6 billion, now all of it fully mapped.

But then, soon after the war, came a bombshell—and it came in the shape of yet another, and equally ambitious series of maps of the surface of the world.

All of a sudden, commercial aviation blossomed from an entertaining curiosity into a formidably important component of the economy, for the rapid transport of both goods and people. And while a map of the world's surface made for the convenience of a surface traveler, a similar map designed specifically for the aircraft pilot became, all of a sudden, an essential. And it did so initially for one very simple reason: a pilot flying two miles high above France had an absolute need to know if there were any mountains on his route that were three miles high, and which he would otherwise plough into. A map of the world's surface that would inform him of such peril would be just as essential for his cockpit glove box as the joystick or the compass or the landing gear.

The International Civil Aviation Organization, initially made up of fifty-four nations with aviation ambitions, was set up in 1947. One of ICAO's priorities was to create a series of charts of the world's surface—stripped of many of the finer and more granular details that would be of little interest to those sitting so high in the sky—but to do so at the same convenient scale as the IMW, one to a million. And in each square of the map, to note in large and very visible figures the height of the tallest object—a mountain, invariably—that a plane might be likely to encounter. One by one, countries that had set their mapmakers to work producing elegant million-scale land surface maps now found they were compelled by international agreements to prepare brand-new aviation charts of this kind, similar to but different from the IMW, and, because of the demands of airlines, to do so at a dramatically faster pace.

Qantas and British Overseas Airways Corporation (BOAC), for example, both began long-haul services between London and Sydney in 1947—with refueling stops in Rome, Tripoli, Cairo, Karachi, Calcutta, Singapore, and Darwin. The cockpits of aircraft flying the route—which took the aircraft fully four days, before jet aircraft replaced the lumbering Lockheed Constellations—thus had to have a quiverful of up-to-date physical outline maps of Europe, the Levant, North Africa, Iran, Afghanistan, India, Burma, Thailand, Malaya, the Dutch East Indies, and all of Australia from the Gulf of Carpentaria to Bondi Beach. The agencies charged with either making these new maps or providing information for others to make them had little time to produce at the same scale the much more detailed IMW maps of what were, one would have to confess, of lesser immediate importance.

And so the IMW, now subsidiary in importance, started to wither and to die. The United Nations took over the project and tried to inject some enthusiasm, but to little avail. By the 1960s only about four hundred of the sheets were fully finished and released (many countries suddenly got cold feet about offering the maps to the project, even though they had been completed), and though each of the finished sheets had the appearance of great elegance, most were out-of-date by the moment they hit the streets.

To incorporate all of the 37 billion acres of the world's land surface into the project would require more than a thousand sheets. By contrast, the ICAO aviation maps, shorn of all but the most vital information and so much simpler to make, extended across all the land surface, all the myriad islands in mid-ocean, and all the oceans, too. They were eventually to be named the World Aeronautical Charts. Maybe they were a little wanting of cartographic elegance, but they were easy to bring up-to-date and presented all the essentials, and turned out to be wholly fit for purpose in a way that their hapless IMW cousins were clearly not.

Albrecht Penck's dream of 1891 finally submitted to its melancholy and officially euthanized end in mid-December 1986. That was when a U.N. "ad hoc group of experts" met in Bangkok, declared that the IMW was of no further value or need, and formally let member states know they need devote no more time or energy to producing it. A complete set of their combined efforts was handed for posterity to the American Geographical Society, which curated the sheets for some years at its various grand headquarters in Manhattan; but when the society—not as flashy and well funded as the National Geographic Society in Washington, and suffering from the current popular indifference to geography— moved to a more modest home in Brooklyn, the maps were transferred to the all-encompassing cartographic library at the University of Wisconsin's campus in Milwaukee, available to be seen and marveled at by anyone wishing and willing to make the trek to the upper west bank of Lake Michigan.

Their curator, proud to preside over such a collection, slides out the hundreds of virgin sheets one by one from their rows of steel cabinets. And yes, they do indeed present a wondrously artistic vision of the world's solid surface. But they show only a fraction of it, by no means the world's totality. That it took the combined and concerted efforts of mankind so long, the better part of a working century, to make a portrait of its real estate holdings that is so circumscribed suggests many things: laziness, indifference, a lack of political will, the sheer complexity of the subject, the logistical cat-herding nightmare of persuading so many rivals to set differences aside in the interests of planetary communion.

Much the same trials afflict a similar, more recently commenced effort to map yet another land surface— the submerged underneath of the oceans, the landscape of the seabed. Such charts as we have made thus far show the details of the seabed to a resolution several orders of magnitude less than those satellites have allowed us to produce of the surface of the moon or of Mars. So a concerted attempt, principally financed by the Japanese Nippon Foundation, is now under way to produce a general bathymetric chart of the world ocean, such that in time the whole solid surface of our planet is mapped, whether it be exposed or submarine, and we have a knowledge of the entirety of our planet, every acre of it—the 37 billion acres that are exposed, or the 90 billion more that languish beneath the sea. It remains a vision, a dream, a hope.

4

At the Edges of Worlds

The two frontier towns are less than a cannon-shot distant, and yet their people hold no communion. The Hungarian on the north, and the Turk and the Servian on the southern side of the Save are as much asunder as though there were fifty broad provinces that lay in the path between them. Of the men that bustled around me in the streets of Semlin there was not, perhaps, one who had ever gone down to look upon the stranger race dwelling under the walls of that opposite castle.

—ALEXANDER KINGLAKE,
Eothen (1844)

There is a zone of twenty precisely measured feet of naked North American land, kept strictly clear of people and trees and all invasive botany and on which no new buildings may ever, without permission, be erected nor any ditch dug, nor any road or bridge built, nor any pipe or electrical cable established, and in the exact middle of which stand a large number of white granite markers, a line connecting which officially demarcates the mutually agreed boundary that separates the Dominion of Canada from the Republic of the United States of America.

This is an international land border—at 5,525 miles the longest in the world. There is currently no wall or fence or watchtower built on or along or across the line, except for those movable gates or barriers that have been constructed at the 105 roadways where it is legal to pass through the border, and at the ends of the fourteen tunnels or bridges where the boundary runs along a river or across a lake. It is said, proudly but these days quite erroneously, to be *undefended* and thus the longest such *undefended border* on the planet. The line that demarcates the two neighboring states has come about as the result of a number of complicated negotiations and subsequent agreements that have been signed during a lengthy period from the late eighteenth until early twentieth centuries by plenipotentiaries initially from the courts in such places as London and St. Petersburg, and with the signing ceremonies conducted in variety of cities, most of them in Europe, but with the American capital city of Washington, D.C., party to nearly all of them.

By the latest count there are 316 other international land borders around the world. These divisions extend—usually in ragged fashion, and seldom in the kind of straight lines that bound so much of the United States—for something like 154,000 miles. Some of those miles are marked by natural division lines (rivers, mountain chains, swamps) while some others are lines drawn almost at random by politicians and civil servants. There are also 193 island-nations in current existence, but these do not by their very nature have land borders, except for those—Haiti, for example, St. Martin,* Papua New Guinea, and Ireland among others—that sport such a boundary within their own maritime territorial fastnesses.

---

* The Caribbean island of St. Martin—known also as Sint Maarten—enjoys the unique position of being the relict colonial possession of two European powers, such that the flags of France and the Netherlands fly at its internal land border. But since both powers are now members of the European Union there are no customs checks. Unlike St. Martin, which employs the euro, Sint Maarten sticks to the Caribbean guilder, for inexplicable reasons.

Each one of the world's 317 land borders is defined and agreed to, formalized as a result of agreements between those countries that abut each other, as neighbors; and each has been surveyed with, in most cases, a mutually agreed level of precision. To read a table of these boundaries and their history is to learn much, in the same way that stamp collecting, now in popular disfavor, used to teach much about the world's historical geography. One can derive great pleasure in picking at random from the United Nations list: learning of how, for instance, the Albanian border with Montenegro was first agreed to by a delegation of Turkish pashas who went to Germany and signed the Treaty of Berlin in 1878; that the northwestern border of Myanmar, separating it from the Indian state of Manipur, was the result of a victory over the Arakanese by the Burmese army back in 1558; that in 1821 an entity called the United Kingdom of Portugal, Brazil and the Algarves annexed a free-spirited confection of small states gathered around the Río de la Plata's estuary and known at the time as the Liga Federal, and in doing so set up a line still recognized today as the border between the modern countries of Argentina and Uruguay; and that the border between Ireland and Britain came about in 1921 when the twenty-six southern counties of the Irish Republic, all with a majority population of

Roman Catholics, declared an independence from Britain which the six mostly Protestant-majority counties of the northeast of the island could not and would not accept, and who thus remained loyal, if troubled, and protected behind what would become a highly militarized border, for much of the century beyond.

Some borders are of great antiquity. As to which was the first—there are many candidates. There was probably a time during the formation of early civilizations when, say, that in the Nile Valley expanded outward like penicillin in a petri dish, and collided, as it were, with its Persian equivalent spreading outward from the Fertile Crescent. The point of close encounter of these two spreading human populations could well have been near where stand today's cities of Basra and Abadan, the waterlands of the Marsh Arabs, as well as the site of ancient Ur of the Chaldees. But no archaeological remnants of a formal border has ever been found in the local estuarine sands—just oil fields and argument, the common coin of this much despoiled corner of the world.

It is said by the people of Andorra that their seventy-mile long Pyrenean frontier with France is the oldest established such entity in the world, having been jointly agreed, written into a signed charter, and then marked out with boulders on September 8, 1278. Are there contenders still more venerable than this?

Hadrian's Wall, for example, which the Roman legions scythed across the windblown fields of the far north of England, was begun more than a thousand years earlier, and can fairly be said to have been at the least a precursor of England's present-day border with Scotland. The Great Wall of China is mainly a fourteenth-century creation (though some of its sections were built fifteen hundred years earlier) and it can be seen as a national border, too. There is an important difference, though. In Andorra, rulers on both sides of the proposed border agreed to its delineation; but neither Hadrian nor the Ming wall builders ever won the agreement of the inhabitants beyond their constructions, who never signed off on the notion that this should be a jointly accepted and formal boundary. The line was unilaterally imposed rather than mutually accepted—a difference suggesting that the Andorran claim to primacy is most probably reasonable and legitimate.

The great majority of the world's land borders were fashioned in the late nineteenth and early twentieth centuries: a fierce acceleration of nation building got under way in 1850, became territorial mayhem between 1875 and 1899—largely in consequence of European adventurism and imperialism—and reached its climacteric in the first two decades of the twentieth century, when 50,000 miles of extra borders, almost a third of those currently inscribed in our atlases, were agreed to and delineated. Many of them were components of the new postwar world order that was drawn up by the Paris Peace Conference of 1919. A flurry of new boundary-making etched its imprint onto maps and charts after the Second World War, and a fair additional number of lines were added after the Balkan conflicts remade such ancient entities as Kosovo, Bosnia, and North Macedonia where once there had been more simply the single republic of Yugoslavia. But then, however, the territory-making energies became largely dissipated, and now most of the world's borderlines are more or less fixed and stable. Maybe the names of the countries they enfold change from time to time, and the politics within often alters on a whim. But to all of this the land itself remain sturdily indifferent and unmoved, the human behavior played out on its surfaces merely trivia. Except, of course, where human behavior induces changes to the ferocity of the weather and the levels of the sea, and land may then fall victim to the climate, and has to alter its shape and size as a result.

Yet the boundaries of all too many of today's countries have less to do with nature—as with China and India, separated by the Himalayan range, or as with Chile and Argentina, divided by the Andes—than they do with the seemingly random apportionment of the hinterlands by politicians, generals, or faraway officials. The United States is one such, its borders being in large part geometric—a straight line drawn for many hundreds of miles and having little or nothing to do with any physical need for separation. A section of its border with Canada, the hundred-mile reach of the St. Lawrence River in upstate New York across from the Ontario towns of Kingston and Cornwall, runs along what is an actual physical barrier, with the centerline drawn in midriver above the thalweg, the deepest point of the stream. Four of the five Great Lakes can similarly be said to offer natural barriers, but the borderlines themselves have been drawn across Lakes Superior, Huron, Erie, and Ontario in straight and decidedly non-natural ways, paying no heed to topography at all.

The situation of the Republic of India is even more spectacularly detached from physical reality, since its borders on both its western and eastern sides were determined by nothing more than the exacting cartographic effort of a civil servant based in London, and who drew the lines on a map in a scant seven weeks in the high summer 1947. His work, resulting in a sudden and near arbitrary division of the countrysides of the Punjab in the west and Bengal in the east, led to an orgy of killings and rapine quite unanticipated in its scale and ferocity, and never to be forgotten.

The man proximately responsible for the tragedy of India's partition was Sir Cyril Radcliffe, a bespectacled and mild-mannered Welsh lawyer who had never before been to India—nor ever traveled east of Paris, indeed—when he got the call. His instructions were as brutal as they were simple: break India apart on religious grounds that would satisfy the demands of the country's wily Muslim leader, Mohammed Ali Jinnah, for a separate state. The demand for Radcliffe to do the job—made because of his repute as director-general of Britain's Ministry of Information—came from the figure whom history has more justly blamed for the debacle, and who was the ultimate decision maker: Louis Mountbatten, the final viceroy of India.

Lord Louis—handsome and debonair, spangled with more medals than wars were ever fought, bathed during his long lifetime by a steadily increasing cascade of public reverence—enjoyed a career that took him from being a purported naval hero (played, barely disguised, in a laudatory film by Noël Coward), to a victorious wartime commander in chief in the war with Japan, to viceroy of India and, besides, the uncle of and adviser to Prince Charles, the future British sovereign. But then he was assassinated at the hands of the Irish Republican Army in 1979, and ever since his reputation has swiftly tumbled, not least because of the heartless manner in which he broke India apart, and how he presided without much evident distress over the consequent rioting and rampage which took a million lives, at the very least. His repute is now in such tatters that when an Indian filmmaker looked closely at the manner in which Radcliffe, as head of the two boundary commissions that drew the dividing lines, actually partitioned the country, the genteel British actress who played Lady Radcliffe remarked that Mountbatten was thought of so comprehensively crooked that "if he swallowed a nail he'd shit out a corkscrew."

The film's young Indian director, Ram Madhvani, made his ten-minute dramatized documentary to depict a 1966 drawing room in Warwickshire, with a retired and nearly blind Radcliffe first being telephoned by the BBC to be told that W. H. Auden, no less, had just written a poem about him and his attempt—despite being neither a cartographer nor an administrator, or in any way familiar with India—to pry apart the countries.

He is at first flattered by the sudden attention, the sudden jolt to his otherwise peaceful retirement. But then his wife, bent on consoling him, says that the poem is anything but flattering, and in the brief and unbearably poignant film she proceeds to read it to him, allowing the viewer to see a decent man being slowly shattered by a memory of the terrible effects

Sir Cyril Radcliffe, an eminent British lawyer who had never been east of Paris, was the eccentric choice to draw the boundaries of partitioned India. The orgies of killing that resulted shattered him. He refused all pay and burned his notes.

of his having so broken apart a hitherto unfractured land.

Unbiased at least he was when he arrived on his
  mission,
Having never set eyes on this land he was called to
  partition
Between two peoples fanatically at odds,
With their different diets and incompatible gods.
"Time," they had briefed him in London, "is short.
  It's too late

For mutual reconciliation or rational debate:
The only solution now lies in separation.
The Viceroy thinks, as you will see from his letter,
That the less you are seen in his company the
    better,
So we've arranged to provide you with other
    accommodation.
We can give you four judges, two Moslem and two
    Hindu,
To consult with, but the final decision must rest
    with you."

Shut up in a lonely mansion, with police night
    and day
Patrolling the gardens to keep assassins away,
He got down to work, to the task of settling
    the fate
Of millions. The maps at his disposal were out
    of date
And the Census Returns almost certainly incorrect,
But there was no time to check them, no time to
    inspect
Contested areas. The weather was frightfully hot,
And a bout of dysentery kept him constantly on
    the trot,

But in seven weeks it was done, the frontiers
    decided,
A continent for better or worse divided.

The next day he sailed for England, where he
    quickly forgot
The case, as a good lawyer must. Return he
    would not,
Afraid, as he told his Club, that he might get shot.

The two boundary lines that Radcliffe and his team of mapmakers drew, working on the dining tables beneath the ever whirling fans of the old deodar-wood house up in the cool of Simla, snaked with passionless authority through the wheat fields of the Punjab and the rice paddies of Bengal—creating a Muslim-dominated West Pakistan and a similarly Muslim-majority East Pakistan. Immediately after the borders were announced on August 17, 1947—two days following the declared independence of India and the formal lapse of British rule—surges of panicked millions began, in a mad and barely controllable rush of humanity, to reach those countries to which their religions now said they belonged. Muslims in Lucknow fled wildly to the presumed safety of Pakistan; Hindus and Sikhs trapped in

Lahore battled to get across to Delhi and Amritsar and the supposed sanctuary of India. It was an unutterably and months-long three-way exodus, quite terrible in its scale and bloodshed, quite shameful for the British who first set the process in motion, and with the apportionment of consequent blame justified, widespread, and near endless.

Sir Cyril Radcliffe himself left India the day after his work was done; he burned all of his papers and, in dismay and disgust, returned the fee of some five thousand dollars that he had been promised. Most biographical mentions of him minimize mention of his work in the subcontinent, even though it remains the Radcliffe Line—This Bloody Line, he called it—for which he is best known.

It may be easy to blame the outrage over the drawing of the Radcliffe Line on the plain fact that it was British inspired, its twists and turns drawn with a British fountain pen by one of the British empire's architects. But that would be too simplistic. Over at the Khyber Pass the Durand Line, also a relic of the empire, drawn half a century before, confirms a natural division between empires and civilizations that is in most places marked organically, by ranges of mountains.

It is notable that the Durand line had been formally surveyed and marked by a succession of joint Afghan-

British teams, working under the superintendence of the British civil servant after which the line is named, Sir Mortimer Durand. Moreover, the positioning of the marker stones—of which there are hundreds, found today in some of the wildest and most lonely landscapes in the world—was agreed to by Afghans and Britons working together. And although the current Afghan government rejects the line's legitimacy, and while many local tribes through whose lands it has occasionally sliced in an arbitrary manner squabble over it as well, it has remained stable and intact for more than a century. Since by and large it confirmed most existing ethnic divisions, it never quite excited the kind of passions that enfolded the Radcliffe Line some few hundred miles away to the east. In summary: the Durand Line enjoys a stability because it reflects, by and large, an existing physical reality. The Radcliffe Line, by contrast, defines only the randomly and hurriedly decided pair of eastern and western borders of today's India.

Insofar as both these latter borders are artificial, made along lines drawn along historically administrative rather than physical divisions, drawn with unreasonable swiftness and cavalier thoughtlessness, and contrived by a foreign-decided partition and thus in no sense as a reflection of any settled order of local history or geography, they are dangerously unstable—as most of the world's partitions are. Sir Cyril never spoke in detail of what he had done, steered all conversations away from his eponymous creation—achieved, for better or worse, by his patiently examining old survey maps, drawing lines as best he could through regions which out-of-date statistical data suggested were majority-this or majority-that, and by doing so helping, if inadvertently, to create mayhem as he went and a legacy of bitterness and anger and on occasion, farce, which continues to this day.

As mentioned above, there is effectively only one border crossing on the India-Pakistan border today, only one formal breach in the barbed-wire jungle that now marches along Radcliffe's elegantly drawn line from the Rann of Kutch up to the Himalayan snow-fields. On large-scale maps the opening stands between the two large cities of Lahore and Amritsar; on more detailed charts, the border is breached between a pair of nondescript Punjabi villages, five miles apart—the one named Attari, which ended up on the Indian side; the other Wagah, which is now Pakistan. The crossing place is generally now known as Attari.

Attari used to be an easy enough crossing—providing, when you were leaving India, there was no alcohol in the car. In the early days one would import half a dozen bottles of whisky to give to Pakistani friends, by dint of giving one of them to the customs officers. That all changed in the late 1970s, when it became well-nigh impossible to get any strong drink across the frontier. The bottles were then either confiscated or smashed.

Today the border is a much more fiercely guarded affair, all-but impenetrable. It is something of a challenge to remember that this demarcation of land in the Punjab should in principle be very little different

The evening gate-closing ceremony at Wagah, the only crossing-point between India and Pakistan, has assumed the role of a bizarre theatrical performance, with soldiers from each side trying to impress rival crowds and whip them up into frenzies of mutual hostility.

from the separation of the furrows and fields back in Deverel-Rimbury England, four thousand years before. Farmers had drawn furrows in the Punjab for centuries, and boundaries between their farms, between their villages, between their towns, had once been informal affairs. But here and now, politics has intruded. As has empire, and geopolitics, and argument, and, on many occasions, war. Very much more is at stake in this imperially delineated stretch of land than ever was the rivalry between wheat- and einkorn-growing farmers. Those on either side of the line between Attari and Wagah today are heavily armed, and with an arsenal of atomic weapons, too. The land border here has the future of much of humankind enfolded into the security of its own existence.

And so there are huge numbers of heavily militarized sentries—Pakistan Rangers and the Indian Border Security Force on each side of the double rows of razor wire, with thousands of mines planted in the space between the fences. There are powerful arc lights along the entire length of this national border, from Kutch to Kashmir, making the line entirely visible from the International Space Station, a thin line of orange tracing its way between the glows of the big cities of the Punjab. From space it mimics Cyril Radcliffe's fountain-pen line, tracing its black ink across the coordinates of his inexact map. The land beyond the cities is black in the NASA images, especially on the Pakistani side of the fence: the farms of the Indian Punjab, so often managed by Sikhs, are rather more productive, and they have electricity, in the main, giving the land on the Indian side a faint glimmer of illumination, an impression of the existence of rather more economic vibrancy here than in the fields around Lahore.

The ideological chasm and the deep political differences and religious hostility between the two neighbor-nations have been turned into a bizarre theatrical performance, staged each evening with a formal ceremony that culminates in the slamming shut of the iron border gates. Soldiers from both sides, selected for their height and their balletic marching skills and ability to kick their legs high into the air, march fiercely toward one another, stopping fast at the painted line on the road so they are almost moustache to moustache, breathing hard and angrily into each other's faces, while the crowds on bleachers on each side cheer lustily and, just being able to see one another, roar insults into the gloaming. Flags are lowered, inch by inch, gates are opened and closed and opened again, until, with a final perfunctory handshake across the line—the Radcliffe Line, *This Bloody Line*—that suggests the whole thing is a grand joke, the gates are banged shut and locked, the soldiers march away and the crowds go back to their mosques and temples and gurdwaras in Lahore and Amritsar and Wagah and Attari, and the border falls quiet, just a line of orange lights and a million land mines and thousands of soldiers on perpetual alert, waiting for the next outbreak of unpleasantness along a line said to be even more dangerous—as President Clinton once remarked—than any other border in the world.

A thousand miles to the east, across in the northern part of Bengal, Sir Cyril was active too, and so there is yet another set of fences and there are lights and more soldiers and all the usual paraphernalia of this unhappy division of British India's myriad spoils. But there is also the existence, or its relics, of a curious cartographic eccentricity—a large number of complicated parcels of land that are defined as *exclaves*, in which until recently tens of thousands of people were stranded and trapped inside what in technical were islets of countries not their own, with their inhabitants never able to get out and return home.

This is land demarcation made insane. It has all to do with the existence of the once princely state of Cooch Behar—landlocked and nestling to the south of Bhutan, to the northwest of the tea-growing countryside of Meghalaya, and close to the slow-flowing and miles-wide Brahmaputra river. Hindus, Muslims, and Buddhists all contested or passed through what was a strategically significant land corridor here, a passage that linked India to Burma and beyond. In the early eighteenth century a large number of clashes broke out between Mughal rulers on the one hand—this was before the British had seized India, and when the Mughals were in the dying days of their rule—and on the other hand the soldiers of the maharajah in Cooch Behar. At the conclusion of each skirmish, a peace treaty was signed, giving this village to the Mughals, that village to the maharajah, and other smaller pieces of land to the various chieftains who served at the maharajah's pleasure. It happened again and again. Every fight ended with a treaty; every treaty apportioned some land to a winner, some to a loser. The consequences of three centuries worth of defiance, magnanimity, and goodwill left a cartographic and humanitarian nightmare, a patchwork of staggering administrative complexity. And all set down among a landscape, with rice paddies and hills of the greenest beauty and loneliness, like nowhere else on the planet.

And that was merely in Mughal times. As this part of India became subsequently ruled by the British, by the local princely states, and then sliced asunder by Cyril Radcliffe's fountain pen to be divided between Indian

Bengal and by the short-lived entity of East Pakistan, which became the new nation of Bangladesh, so this strange complexity of broken-off morsels of states remained, endured, and became notorious for its singular *oddness.*

Thus, here and there in tiny dots of historical bizarrerie, inside tracts of territory that were indubitably Bangladeshi, was a marooned flyspeck of land that, because of a two-century-old peace treaty, was still actually a part of India. And similarly, in dozens of instances, small and orphaned portions of Bangladesh were in fact swimming inside the immensity of some of India's sovereign borderlands. To make matters even more weird, even tinier bits of India—with people, hundreds of people in some cases—lived inside some of these minuscule parcels of Bangladeshi territory that were themselves inside India. And in one particularly and unutterably mad instance, a village of undoubted Indian ownership was inside a piece of Bangladeshi territory that was itself inside an Indian parcel that has been somehow pinioned inside Bangladesh. Four mutually enclosed levels of sovereignty, getting progressively smaller like *matryoshka* dolls.

The borderlands in these few hundred square miles contained some hundreds of enclaves and exclaves with tens of thousands of people who—for example—could not claim their citizenship, if they were Indian, say, because to visit an Indian consulate obliged them to pass through Bangladeshi territory—*for which they did not have a visa.* The Indian government, its local officials high-caste *babus* seemingly schooled in the minutiae of pedantic exactitude, was especially harsh in enforcing its own immigration rules, and it was said that some 75 percent of those who lived in Bangladesh enclaves spent time in Indian prisons for having—either deliberately or inadvertently—passed into the Indian territory into which history had so pitilessly enfolded them.

It would be unfair to suggest that here was a place where the land had gone mad. It would seem proper to suggest, however, that here was a place where the men that ruled the land had gone mad. The land is of course always indifferent to the insanity of those who live on it; but here one imagines that the patience of the acres must have been sorely tested.

Gradually, it appears, political will and maturity (though of a peculiarly subcontinental kind) seem to be groping their way toward resolving the situation. Proposals are currently being put forward for enclaves like these to be formally swapped between India and Bangladesh, though it is all happening with much bureaucratically frustrated and characteristically subcontinental tardiness.

The first attempt to do so, wholesale, was in 1958; another try, based on legislation written in faraway Delhi, was made in 1974; and a third was attempted in 2011. And still enclaves and their enclosed people remain—in 2015, some 38,521 people lived in 111 Indian enclaves inside Bangladesh and 14,863 Bangladeshi villagers lived in 51 enclaves trapped inside India. Although both governments insist that progress is being made to solve the problem of these manifestly unlucky people, it does so at a lumbering, elephantine pace. To these unfortunates, held in a time warp of Kafkaesque dimensions, the thought that a civil servant from London could once have come and established on the far side of the country an impenetrable, indelible, and still permanent borderline in just thirty-six days in the summer of 1947, while their landlocked imprisonment derives from the eighteenth-century history of the maharajahs of Cooch Behar, simply beggars belief.

Yet India does not have a monopoly on such a self-evident borderline personality disorder. There are oddities with America's northern frontier, too.

Unlike the more troublesome southern border with Mexico, the frontier with Canada has a general appearance of geometric tidiness. For some 2,175 of its miles it follows the 49th parallel, straight as a die aside from a few hundred-foot deviations made inadvertently—intense cold, and strong drink, may have been involved—by those nineteenth-century surveyors charged with laying it out. The very different, topographically much scruffier part of the borderline in northern New England is home to many of the better-known eccentricities—the library in Vermont with the frontier passing through its lending room, a hotel similarly divided between Quebec and New York state, and a number of houses split in two by the unyield-

The thousands of miles of supposedly unguarded—but electronically monitored—frontier between Canada and the United States is marked by a twenty-foot permanently cleared vista to help define and police the borderline.

ing demands of treaties and surveys in the nineteenth century. Time was when both countries regarded these oddities as amusing; but the U.S. Border Patrol officers nowadays—perhaps understandably—have had something of a sense-of-humor failure and tell passersby quite forcefully that, for instance, they may use only the Canadian sidewalk in the settlement of Derby Line, Vermont, since the pavement itself is U.S. territory, and merely to stroll on it without a visa is a violation of immigration law, and can result in a fine or worse.

There is one glaring—and similarly rather strictly policed—oddity at the very eastern end of the 49th parallel. A strange historical error—followed by the stubborn intransigence of the new U.S. government— has resulted in the existence today of a somewhat inconveniently sited American enclave, 123 square miles of it, and which is only accessible, if trying to get there by land, by way of passing through Canada. It is called the Northwest Angle, it is the most northerly point of the contiguous forty-eight states of the United States— and it derives its existence from a simple mistake made in the early nineteenth century by explorers who believed, wrongly, that hills to the west of the Lake of the Woods were the source of the Mississippi River.

The Lake of the Woods, a wildly irregularly shaped body of water that straddles Ontario and Manitoba, and provides Winnipeg with much of its water supply, is a relic of the once massive Lake Agassiz, one of the largest-ever glacial lakes in history. When the immense volume of Agassiz freshwater finally drained out through a broken ice dam, eight thousand years ago, the world's sea levels steadily rose by as much as nine feet—an event that gave rise, many suppose, to the biblical myths of the Great Flood, Noah's Ark, and the supposed strandings on the summit of Armenia's Mount Ararat. Fanciful these myths may be, but the legacy of the Lake of the Woods remains, and its very existence helped create a geographical fancy of its own, two centuries ago.

The origins of the error go back to the Treaty of Paris of 1783, which fully recognized the legitimate existence of the new-made republic of the United States, and granted to this new country perpetual title to its immense extent of lands. Exactly what these lands comprised and where they began and ended was a little vague, especially in the north around the Great Lakes—but the treaty's framers, with a small army of geographic advisers, did their best to define the division, the ultimate border, between the latitudinal limits of this new nation and the British-ruled remnants of land to its north, which would become Canada. Until the Louisiana Purchase of 1803, the lands west of the Mississippi River belonged variously to Spain and France, and so the river was considered to be the western frontier. In 1883 it was erroneously assumed that the lake, according to the celebrated British map-maker John Mitchell,* was shaped like an egg, not the wild mess of inlets and arms and outlets that survey-ors later determined it to be. The border would thus pass from "the northwesternmost point of this lake," as the framers of the 1783 treaty wrote, and more critically, "thence across to the Mississippi in the west." The ragged-looking thing that eventuated would be the frontier-line between new-made American ter-ritory and the possessions of France, after which the line would follow the Mississippi southward all the way down to the Gulf of Mexico. So, a border was labori-ously drawn, and inscribed onto the maps of the day. America had its first official outline.

But then, fifteen years later, along came David Thompson, the Welsh immigrant and former Hud-son's Bay Company trader who became perhaps the

---

* A respected expert on the biology of the opossum, Mitchell was also the man who helped discover that plants have either male or female characteristics, as well as being a cartographer of legendary brilliance whose maps were employed as recently as 1932 in the settlement of boundary negotiations between Canada and the United States.

most notable of all the explorers and mapmakers of the Canadian interior.★ He realized that the Mississippi does not rise anywhere near the Lake of the Woods, but much farther south—and since it therefore does not intersect with the Canadian border, there must now be a significant gap in the line between the two countries. The importance of this realization—which initially concerned more trappers than settlers, as well as a large number of puzzled Native Americans, mainly members of the Ojibwe tribes, who weren't certain which country they now lived in—was much magnified when, in 1803, Thomas Jefferson had the United States buy from France the lands to the west of the Mississippi River. The result of the Louisiana Purchase, as it came to be known, was a massive westward extension of the border, and which was agreed by all to be a geometric border that would pass along the 49th parallel of latitude all the way to the Rocky Mountains (known back then as the Stone Mountains).

★   Though little known beyond Canada, Thompson's name appears in the chorus of the rousing a cappella shanty "Northwest Passage" by Stan Rogers, which many polled Canadians believe should be the country's national anthem. It remains a poignant memory to many older Canadians that Rogers died in a fire aboard an aircraft landing in Cincinnati, where he was due to perform—and doubtless play his best-loved song.

But the untidy and very porous gap between the parallel at the Lake of the Woods persisted—and the still more-or-less new American government, stubborn in its independence, refused to countenance any deviation from the 1783 treaty. That document had specified the border march west from the northwesternmost point of the Lake of the Woods—and after some argument the British—arguing for the shape of the borders of Canada—agreed. Except that no one had the foggiest idea where that northwesternmost point was.

Enter David Thompson once again, in 1824. By this time, it had been mutually accepted that wherever the lake's farthest point was, a line could be drawn due south—not to the Mississippi River, which was now irrelevant, all of it now being in U.S. territory—but to where it intersected with 49th parallel. After which the border would then turn away west and head off in a straight line to the Rockies. Thompson after some trekking through the mosquito-infested forests duly found this northwesternmost point, whence it was properly agreed that this should now serve as the far end of the ragged borderline that went in a more or less easterly direction down and across to Lake Superior and eventually to New England. A short connecting line was then accordingly surveyed to go southward to the 49th parallel—and the United States had its border, finally.

Though its geometric smoothness was now interrupted for all time—the agreement was signed in 1842, after a British astronomer named Johann Ludwig Tiarks formally and very precisely determined the coordinates of Thompson's point, at a place on the lake he called Angle Inlet—the eccentricity of what has come to be known as the Northwest Angle has a certain charm to it. It was fully surveyed during a three-year period starting in 1872—work made necessary by the studied carelessness of the framers of the Treaty of Paris, ninety years before.

As of the last census, 123 American citizens live in the settlement of Angle Inlet, marooned inside Canada, yet residents of the U.S. state of Minnesota.

To get to the settlement by land from elsewhere in Minnesota, the process is irritatingly and amusingly strange. You leave Minnesota through a village named Warroad, cross the border into Manitoba, head due west, and then turn north and back east again for some fifty miles of the most tedious and uninspiring muskeg-like prairie landscape imaginable, then cross into the United States once again and after ten miles within American jurisdiction, there is a small hut by the road, with a videophone. A push of a button connects you with an official somewhere far away, you show your passport over the video link, you are asked a few perfunctory questions, and you are electronically waved into the country.

But woe betide any traveler who does not comply. For hidden cameras will find you and your car within moments, and you will get into trouble, mightily, once you try to return. For here, as all along the 5,525 miles of the *longest undefended border in the world* is an array of unseen and unseeable electronic gadgetry that will do its level best to defend the borderline of the United States from all comers, friend or foe. You may come to Angle Inlet, Minnesota, by land, or by water, or in winter, across the Lake of the Woods by ice. But you will be seen. You will be watched. And your presence will be known. This may look undefended. But defended it most certainly is.

5

Drawing a Distinction

So geographers, in Afric-maps
With savage-pictures fill their gaps;
And o'er uninhabitable downs
Place elephants for want of towns.

—JONATHAN SWIFT,
"On Poetry" (1733)

In the early winter of 1965 I drew a little map—or as good a little map as could be drawn with the limited skills of a twenty-year-old undergraduate—of six square miles of the southern end of the island of Raasay, in the Scottish Hebrides. This was a geological map—my degree depended on it—with Raasay chosen because it is a landscape uniquely blessed with a remarkable array of rocks, both in their type and their age. So for instance, while there is on the island a charming miniature and flat-topped mountain named Dun Caan, made of volcanic basalts thirty million years old, and which commands splendid views of the nearby Isle of Skye, there are also rocks laid down from the sea during Cretaceous times, five times as long ago. The sandstones are rich with seams of iron ore that German prisoners of war were ordered to dig out of mines there in the 1940s.

I lived for a month in an all but roofless cottage close to a settlement called Oskaig, on the island's west coast. The midwinter hours of daylight were few enough, and most of them saw me tramping across the machair and along the coastline and up the hills. Each evening I would cook on a Primus stove, I would read or write up my field notes by candlelight. My equipment was rudimentary: an Estwing geological hammer with one end for thumping and the other for prying, a Brunton prismatic compass and combined clinometer, a small bottle of hydrochloric acid, a tiny pocket magnifying glass—and a map, the most essential, and as it happened, my pride and joy. The physical map on which I would plot such geological information as I might discover, and which I would eventually submit along with some fifteen other examination papers, in the hope of eventually being awarded a degree. (As indeed I did.)

My invaluable possession during that time on Raasay was Sheet 25 of the United Kingdom Ordnance Survey Maps, Seventh Series. It was titled "Portree," this being the largest town on the sheet, although it is in fact on Skye—the largest community on Raasay then having no more than a hundred inhabitants. The map was printed in ten colors, was specially strengthened by being pasted onto folded cotton cloth. Cloth backed, it was priced at five shillings, a shilling less for paper alone. The scale of the map was famously 1:63,360, a ratio familiar to all Britons of my generation. One inch on the map was equivalent to one statute mile on the ground.

There were back then 190 such sheets covering the entirety of the United Kingdom—from the northernmost Sheet 1, "Shetland Islands—Yell and Unst," which included the two most northerly islets of the entire United Kingdom, Muckle Flugga and Out Stack; down south to Sheet 190, "Truro and Falmouth," in southern Cornwall. Today the scales are a little different—with metric measurements replacing the old imperial units of miles and feet—and the much loved 1:63,360 has been replaced by the more bloodless 1:50,000 scale, which now covers the country in 204 sheets—or, if you

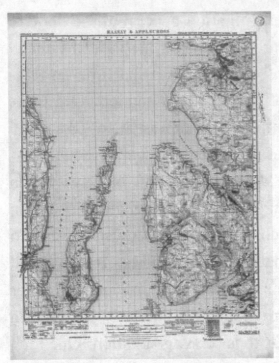

The Hebridean island of Raasay as depicted on a classic Ordnance Survey map of a kind widely revered for two centuries as a nonpareil of the cartographic craft.

want to walk or cycle and need the even larger scale of 1: 25,000, 403 sheets.

But these technical details tend to obscure the fact that Britons, who have had access to such maps since the Ordnance Survey was founded at the end of the eighteenth century, are wholly in love with their OS maps, and cherish them with a fervor that is incomprehensible to most others. No American, so far as I am aware, ever professed a deep and unsullied affection for the USGS topographical sheets that it is possible to order from the government agencies. They are fine enough maps, and they cover the entirety of the nation. But seldom are they bought for the sheer pleasure of ownership, of the ability to pore over them and imagine, or remember, to draw contented admiration at their elegant appearance and scrupulous accuracy.

It is still a rite of every British schoolchild's passage to learn what are known as Conventional Signs—the marks that are common to all the OS maps and which indicate the presence of types of landscape, of highways and byways, of notable buildings of one kind or another. There are particular lines that show roads more than twelve feet wide, subtly different lines indicating roads less sizable. Highway gradients more than 1 in 5 get a pair of chevrons. Military firing ranges are noted, as are post offices, police stations, greenhouses. Windmills are symbolically differentiated from wind pumps, lighthouses get a sign slightly different from those denoting light beacons, sands are drawn differently from mudflats, forests with coniferous trees with different symbols from those wooded with their deciduous kin. Vertical rock faces are marked differently from vertical cliffs. There are conventional signs for boulder fields. Orchards. Roman sites. Non-Roman castles. There are contour lines and, of course, at the summit of hills indicated by a rounding of the contours, there is the tiny triangle with a dot in the middle, representing the triangulation points, the "trig points" as Britons call them, that have made the making of these maps possible in the first place.

And all these things are just, exactly, precisely, where the map says they should be: I puffed my way to the top of Dun Caan, hammer at the ready, magnifying glass primed to inspect the granularity of the Raasay basalt, and there before me, a pillar of concrete rose from the bog and on the top of it three brass rails where once a theodolite had stood, and in the center a brass circle with four words inscribed: Ordnance. Survey. Triangulation. Station. On the map, at this very place: a triangle, a red dot in the center.

No one would think—especially after all the effort of climbing to find a trig pillar—of ever spoiling or interfering with such a memorial, with most regarding them as precious way stations in the country's history. The only joke ever played on the heroic memory of the old surveyors is that visitors sometimes pile a cairn next to the trig point, by doing so making certain that over the years, with all visitors adding a customary stone or two to the cairn, it rises eventually higher than the original point, which then becomes technically demoted, if ever the mountaintop were to be surveyed again.

And inevitably enthusiasts seek to visit all the points in the nation. There are said to be about 6,200 of them, and recent images in the British press show a middle-aged and bearded man, wearing an anorak to keep out the wind and rain, standing proudly by a pillar at the top of a low hill in eastern Scotland. He claimed to have visited every single one of them, which had taken him fifteen years. He said he could only have found them because he owned and consulted every single one of the OS maps in print.

All British bookstores carry the maps, and in areas where there is much countryside, even small general stores will stock them too, familiar in their racks with their pink or orange covers (depending on the scale) and with waterproof versions available for those determined to use them outside, in all weathers.

The maps are cherished both for what they are and, more important, for what they stand for. Their existence says something notable about the notion of freedom, it is generally agreed—the freedom for the ordinary Briton to enjoy a relationship with the land that is unique, and which is defined in particular by a public determination that, even if land is owned privately, everyone should have a right of access to it. The Ordnance Survey maps, so wildly popular, demonstrate exactly how this access might best be gained.

The complicated philosophical and legal questions of ownership and its bastard child, trespass, belong to a later chapter; but insofar as this section has concerned itself with the defining and delimiting of the world's land surface—from the surveying of the world and the marking of the great ethnic frontiers to the matter of being able to stroll the boundaries of the tiniest of parishes—then it is well worth noting how, in some societies, with Britain perhaps in the forefront, the sheer existence of maps of great beauty and simple utility like these have a profound effect on the relationship of the users of such maps with the land that they illustrate and describe.

It is by no means a coincidence that the Ordnance Survey was born in 1791, during the heady days of the Enlightenment, and that it reached its cartographic zenith during the first half of the nineteenth century, which by general agreement is regarded as the Romantic period of British history, the time of Rossetti and Keats, Byron and Shelley and Coleridge. It was the time also of the Industrial Revolution, when science and literature and art held hands and helped together to shape the fabric of a country that, at exactly the same time, was being fully mapped with beauty, accuracy, and scientific care.

People took holidays from work for the first time, and, inspired in part by the paintings and poetry that so lyrically captured the essence of the landscape, they explored, hiked, tramped, climbed, and rambled through and across and up the countryside, exploring the land from the Scottish Highlands to the wilds of Dartmoor, from the Lincolnshire Wolds to the clifftop meadows of Dover. And to help them find their way during this transformative period in their lives, there now were for the first time sumptuously detailed maps, offered in affordable abundance, of enviable exactitude, in a manageable, understandable, and useful scale of one inch to the mile, and in their own way made quite as beautiful as the landscape that they illustrated.

Some philosophers, wrote a later historian of the survey, "feel that maps stimulate free wondering as well as wandering, that they encourage the mind to eschew linear logic and play with randomness and free association, like the rambler who leaves the confines of major roads to skit over wide expanses of purple heath.

"It was during the Enlightenment that maps acquired their associations with political liberty and egalitarianism, and it was during the Romantic period that this was supplemented by a deeply felt love for nature and solitary wandering. Proponents of both movements projected these ideas onto the great 'national undertaking', the Ordnance Survey."

It is these days, and surely rightly so, deeply unfashionable to speak sympathetically or with anything other than contempt and scorn about the British empire. The reasons are obvious, and legion, and millions around the world have good reason for remembering the British with disfavor. "When the missionaries came to Africa," Desmond Tutu famously (though not originally: that honor belongs to Jomo Kenyatta) remarked, "they had the Bible and we had the land. They said 'Let us pray.' We closed our eyes. When we opened them, we had the Bible and they had the land."

But during the heyday of Britain's colonial adventurism, the British did also come to possess some places where there were no resident peoples to oppress or imprison or enslave, where there was no concept of "our land" that might be exchanged for the Bible and a quick prayer. One such, blessed with no native population and possessed by no one, is the island of South Georgia, in the cold South Atlantic waters some few hundred miles off the coast of Antarctica. It is a long, boomerang-shaped island with a spine of sharp snow-covered mountains, scores of glaciers and deeply incised fjords, and with settlements built by late-nineteenth-century Europeans from which they would hunt for whales out in the deep ocean.

It is also an island with an intriguing story on many levels—as the island where Ernest Shackleton and his fellow survivors from the ice wreck of the *Endurance* finally found rescue; as a breeding ground for one of the world's largest seabirds, *Diomedea exulans*, the wandering albatross; as the home of some of the largest king penguin colonies on the planet. And also, most relevant here, as the island where cartographers from England, familiar with the excellence and profound meaning of their homegrown Ordnance Survey sheets, decided during the 1950s to turn their skills to making a comprehensive and wholly accurate map of this very island, eight thousand miles from home.

These surveyor-dreamers were all young men, ex-army in most cases, and, most intriguing, they were led on their expeditions—staged during three austral summer seasons between 1951 and 1957—by a man named Duncan Carse, a figure already known the length of the British Isles, by children especially, for being the voice of the main character in *Dick Barton, Special Agent*, a beloved BBC radio thriller. That a fictional radio detective could be also a skilled mapmaker and a somewhat heroic explorer of the sub-Antarctic—for the ice-bound interior of South Georgia presents a formidably challenging landscape, and there were injuries aplenty on the expeditions—says something both about the temper of the times, and about the nature of the actual workings of empire, the nature that is seen when one looks beyond the cruelties of imperial oppression and intolerance. The science of mapmaking has a certain intellectual honesty to it, its practitioners enjoying a certain independent nobility of spirit. And all of it essential to the demarcation of the world's land, wherever it may be.

PART II

Annals of Acquisition

. . . everyman has a property in his own Person. This no body has any right to but himself. The labour of his body, and the work of his hands, we may say, are properly his. Whatsoever then he removes out of the state that nature hath provided, and left it in, he hath mixed his labour with, and joyned to it something that is his own, and thereby makes it his property. It being by him removed from the common state nature placed it in, it hath by this labour something annexed to it, that excludes the common right of other men.

—JOHN LOCKE,
Second Treatise on Civil Government (1689)

1

Up and Out and
On the Level

The great majority of those in the world today who
lay claim to owning a piece of land have acquired
it secondhand. Generally speaking, land that belongs
to someone now has belonged to someone else before,
and whether it be meadow or moorland, a tract on a
mountainside or a parking lot on Main Street, it can
invariably and rightly be described in just the way that
an old car might be described, or a washing machine:
the land acquired is previously owned, has been lightly
or gently used, is a hand-me-down.

But not quite always. There are of course enormous
acreages around the world that are uninhabitable, or
inaccessible, or seem to be mineralogically worthless,
and so are not presently owned by any individual or
institution and are superintended only by the country

in which they happen to be situated. There are also, however, and thanks mostly to the forces of geology—and in places to the ingenuity and pressing needs of man—small pockets of brand-new land that in most cases never have been owned nor ever will be, and so will never enter the chain of possession that defines almost everywhere else.

Surtsey, off the south coast of Iceland, is one such; it is an island, 600 acres or so of new land, born out of the sea in November 1963 and, as it happens because its rock is loose and friable and easily swept away by waves and wind, slowly diminishing in size. But during its existence so far it has accumulated a fair amount of nonhuman life. There are some fourteen types of birds, with gulls and puffins dominating. There are scores of plants—mosses, lichens, and a hardy bush called a tea-leaved willow. There are seals and limpets, earthworms, slugs, spiders and beetles. But humans are not allowed, neither to visit or to settle. Scientists are the exception, and may go, and there is a temporary bunkhouse from where they may study the development of island biota—though they must live under extremely strict rules, established as a precaution against pollution or contamination.*

---

* The rules are not always followed as assiduously as they might be. One visiting botanist made his morning toilet where

The Kingdom of Tonga in the South Pacific similarly sprouts new islands and did so in the mid-1990s
with a volcano that spurted so much dust into the sky
that several transoceanic airliners had to be diverted.
Once the smoke had cleared, a new island appeared,
smaller than Surtsey but because of its tropical location
burdened with new life much more quickly. The locals
wanted to name their new piece of territory after their
rugby-playing hero, the New Zealander Jonah Lomu,
but he fell ill and died, and instead the island was
given a more traditionally Polynesian name of Hunga
Tonga—and in any case since it, like Surtsey, is eroding and so diminishing in size, to have named it after
a human, heroic or not, would have been symbolically
less than appropriate.

The island known as Anak Krakatau, however, is
expanding in size. This is the child of—*Anak* being
the word in Bahasa Indonesia—the almighty volcano
Krakatoa, which exploded with world-shattering violence in August 1883, and has become the byword as

---

he was not supposed to, and the following season was rewarded
by discovering a tomato plant growing where he had lingered.
Also, a pair of teenage oarsmen stopped at Surtsey, unseen, and
for reasons best known to themselves planted potatoes in the ash
on the beach. All of the offending invasives were dug up and
hurled away before they could take proper hold.

a volcanic superlative ever since. The child-volcano. rising out of the ocean first in 1927, established itself permanently in 1930, half a century after the main explosion, and it did so exactly above the crater of its long demolished parent. The youngster has been growing several feet each year since—it now stands a full quarter mile high, is covered with trees and plants, and sports enormous sea lizards and other spectacular Indonesian flora and fauna. It is also phenomenally active, and its frequent eruptions and lava flows and the occasional regurgitations of boulders as big as London buses pose a considerable threat to visitors. So Anak Krakatau has no permanent human population, nor is it owned by any private individual, but is presided over by the state, which gamely tries to keep all visiting in check.*

---

* White Island in New Zealand, a highly active volcano in the Bay of Plenty in the North Island, and which was in the news in 2019 because a large number of visitors were burned to death in a sudden eruption, is privately owned by a family named Buttle. The major difference between it and the more commonly unowned other volcanoes is that White Island—known as Whakaari to Maoris who first claimed ownership, and were granted it by the Maori Land Court, of which more later—is that local deposits of sulfur could be mined there. There was a commercial reason—and later, and lethally, tourism—for a private individual to want to possess this particular parcel of land.

But nature is not alone in making new land. Land can be created by human agency too, and discrete parcels of new territory have indeed been manufactured, and in a variety of usually overcrowded places, by dint of engineering, by dredging, by excavating and dumping and bulldozing vast amounts of materials, and placing them into the sea to bring new real estate into existence. Hong Kong provides many examples. The Treaty of Nanking in 1841 forced China to cede "Hong Kong Island and Stonecutters Island" to Britain, in perpetuity. Stonecutters, where the Royal Navy used to store ammunition for the Pacific Fleet, in long and deep tunnels carved into the granite, and guarded by fierce-looking Sikhs, magnificent in their blue uniform *pagris,* has not been an island for years. Landfill connects it to the Kowloon mainland now, and on this landfill there are now apartment buildings and railways and subway tunnels coursing through it, and the land of which it is made belongs still to the Hong Kong government, but is leased—as is indeed all of Hong Kong's territory—to those developers who can make money from it.

Likewise, man-made land exists in Manhattan, most notably in the one landfill extension that edges along the all-too-solid schists and gneisses of the island itself. Battery Park City, with its thousands of apartments as well as its hotels, stores, and offices, has been built at the lower southwest tip of the island—all of it on artificial land, which was dredged up in full view of the Statue of Liberty, was stoutly barricaded from the sea and then dried out and tamped down to become such solid land that building proceeded apace for thirty subsequent years—and the land that was made belongs still to a city agency, and no one individual owns any of it.

In San Francisco, too, there is the Marina District—made of landfill also, though perhaps less salubrious than the fresh-from-the-sea sands that were used for most of the extension of New York City. The Marina's basement rock, if rock be the term, was made in part from the building and roadway wreckage of the 1906 earthquake and fire, which devastated most of San Francisco. This provides a level of irony that escapes few who live there and in Battery Park today: both of these highly desirable new-made territories could well liquefy if a seismic event of any significant magnitude occurs. The lands that no one owns will become thixotropically unstable and the buildings erected on them are at risk of plunging directly downward, like rods plunged into pudding.

2
Islands of the Dammed

B ut all of these new lands, whether islands made by nature or extensions made by man, are necessarily small in extent. There is one country, though, where the winning of new land in sizable amounts has become a national pastime, a local industry from which the country wins considerable international repute.

The Kingdom of the Netherlands has spent much of its recent existence manufacturing new territory for itself and the Dutch people—and once having done so, making certain that this new real estate becomes privately owned and with good fortune, turned to personal advantage. Newly made Dutch land, in other words, is officially conceived and its governing policy designed for eventual ownership, to be possessed in the very sense that John Locke had imagined when, two and a half centuries before, he wrote, "he hath mixed his labour with, and joyned to it something that is his own, and thereby makes it his property." Locke supposed—as Thomas Jefferson predicted—such would likely lead to great national prosperity.

The Netherlands is essentially an enormous muddy delta, an amassment of flatness at the mouths of three great European rivers, the Rhine, the Meuse, and the Scheldt. It has been a trading nation like few others, ancient, prosperous, populous, comfortable, respected, and always facing—and always doughtily fighting with—the ever roiling sea. It consists nowadays of twelve provinces, six of which—Friesland, Gelderland, Noord-Holland, Zuid-Holland, Zeeland and, most recently, the improbable comic-book-sounding province named Flevoland—have as a common suffix the word *land*. There is a certain irony here, insofar as a quarter of the country lies below sea level.

The kingdom's livelihood has always depended on its success in fighting battles with the ever pounding—and nowadays, ever rising—waters of the North Sea. To that extent the names of those six provinces may reasonably be thought to have a triumphal tone to them, maybe even hubristically so; or else they may be seen as clarion calls to battle stations, urging their people to keep the will to battle on, to keep the waters away, to keep the sea at bay. Even the country's own name, when expressed in the lingua franca, is a melancholy reminder of the fragility of the existence of its very surface, when it is so threatened by the thunderously close ocean: Koninkrijk der Nederlanden it is called, the Kingdom of the Under Lands. These are the Low Lands. These are the Threatened Lands.

Early settlers in the waterlogged fields of the eastern North Sea did precious little to alter their situation: they lived in and on the bogs, built rafts, bundled rushes, cut peat, and took it as read that the peat holes would fill with water and cause the entire boggy mess they called home to sink further below sea level and thus be flooded and rendered uninhabitable—whence they would move on, in flat-bottomed boats, to marginally higher land nearby and begin again until that excuse for land, too, took on water and slowly foundered.

But then, most probably in the eleventh or twelfth century, came the bright idea of building dikes, wide and strong earthen barriers that would keep the sea back and allow the bog beyond to dry itself out in the summer sunshine, and become usable and habitable and perhaps even permanent land once again. The first such dike was built in the far north of Friesland—twenty-four miles of it, known as the Slachte Dike, and which enclosed an area of marsh between the towns of

Harlingen and Leeuwarden, and which has been protected from the sea ever since. Early on, the means
of ridding a marsh of surplus water were quite basic:
once the dikes were in place so sluices would be built,
with simple lock gates that would be slammed shut by
the pressure of the sea at high tide, but which opened
up once the tide had gone down, allowing any surplus
marsh water to flow outward, and thus make things
drier. It was a primitive one-way valve, and it worked
well enough in Friesland; elsewhere, and as time went
on, more sophisticated methods of sponging out excess
water were devised—not the least of them being the
windmills for which the Netherlands is rightly famous.
Chains of such drainage mills, all employing the stiff
North Sea breezes to lift water from the muddy canals built through the swamps, turned out to be highly
effective—and one by one, tracts of land, none of them
bigger than a handful of meadows worth, were rescued
from the sea. Cows and pigs, tulips and people, swiftly
replaced the peat, cotton grass, and marsh iris that had
hitherto dominated the Dutch landscape.*

---

* The story of a young boy heroically plugging a dyke with his
finger, and thus averting catastrophe, is not a Dutch story, and is
little known in the Low Countries—though there are some statues to the fictional youngster, whose name was Hans Brinker.
The story was written by an American children's author, Mary

The land thus rescued and reclaimed was given a generic name that has since spread worldwide: the polder. The etymology of this word, which has been in the language since the fourteenth century, is uncertain, though it seems to owe something to an earlier Dutch word that gives the English *pool*, signifying a small body of standing water. The polder system has itself moved well beyond the confines of hydrology and surface making: the fact that the making and preservation of a polder requires a mixture of official direction and local cooperation—and all of it strictly and defiantly nonpolitical—has allowed the word to underpin much of the manner in which Dutch society believes it comports itself: collectively, firmly, cooperatively, in a fashion both secular and apolitical, and for the good of the nation and for the people as a whole. The Dutch national philosophy is related to the polder system—unarguably making the Netherlands the only nation on earth with its guiding principles based on the theory and practice of the making of land.

Nature, of course, didn't always allow the polders

Mapes Dodge, who had an aptitude for persuading top-drawer writers to contribute to her monthly *St. Nicholas Magazine*. Ms. Dodge, who died in 1905, is said to have first commissioned Rudyard Kipling to write *The Jungle Book*.

to enjoy a sense of permanence—even though James Watt's late-eighteenth-century invention of the steam engine allowed for the construction of a large number of pumping stations, far more efficient and effective than the windmills and the Archimedes'-screw lifting systems that had been used before. But huge storms and their attendant floods more than occasionally reversed all of even this great progress, and most Dutch people today know all too well the dates of the worst of them— this being a country of long water-related memories. Most will still be able to spell out the dates of the Great Storm of 1703—which Daniel Defoe described as "the most terrible storm the world ever saw"—as well as the Christmas Flood of 1717, the two St. Elizabeth's Floods (of 1404 and 1421), the two All Saints' Floods (of 1170 and 1570), and any number of other massive pulses of seawater named for the feast days on which they first occurred. More recently, the North Sea Flood of 1953 killed 2500 Dutch citizens, and television films of the catastrophe demonstrated to the world for the first time the enormous struggles that the Netherlands have endured, keeping their country intact and afloat.*

★ The experience and ingenuity accumulated by Dutch flood-control and land-making engineers has led to the spread of polder systems around the world: swamplands in England and

The country's dry-land balance sheet, in part a consequence of the storms, was less than impressive. During the seven centuries since they first began building polders, the Dutch managed to reclaim a little less than a million acres of land—but at the same time they lost 1.4 million acres of already solid surface. Most of this had been chewed away by the steady expansion of the enormous and violently rough inlet of the North Sea that dominated the country's interior like a Florida sinkhole, and was somewhat misleadingly called the Zuider Zee, the South Sea. Misleading because it is in the north of the country: it turns out that it was first named by the people of Friesland, where the channels are situated through which the gales and tides push the North Sea waters—and to the people of Friesland, this enormous and tempestuous body of water was indeed well to their south. That it was north of everything else in the country was of no consequences to the somewhat isolated and provincial folk of Friesland.

No challenge was greater in all of Dutch history than the decision, taken over a period of some thirty years at the beginning of the twentieth century, to

---

Russia, Surinam and Thailand, the Pontine Marshes in Italy, and the boggy country near the mouth of the Gironde have all been drained and dried out by ingenious Dutchmen.

bring the balance sheet into equilibrium—by drying out and making polders of the nearly two thousand square miles—1.2 million acres—of the entirety of the Zuider Zee.

Which is where the figure of Cornelis Lely comes in. The famously droll remark that "God made the world—but the Dutch made the Netherlands" could best be applied to Mr. Lely, who is rightly remembered—with honors and much fine statuary—as the principal architect of one of the most impressive engineering projects undertaken anywhere in modern memory. The Ameri-

The celebrated hydraulic engineer Cornelis Lely devised and oversaw the construction of the great dam system that raised from the ocean vast acreages of brand-new Dutch land, essentially also allowing the Netherlands to survive encroachment from the sea.

can Society of Civil Engineers declared in 1994 that the Zuider Zee Works, as the project was formally named, should join the Empire State Building, the Golden Gate Bridge, and the Panama Canal as one of the Seven Wonders of the Modern World.

Lely, who combined his skills as a hydraulic engineer with his position as a Liberal Union party politician, had campaigned since the 1880s for the Zuider Zee to be reclaimed. In 1893 he came up with a formal engineering plan: there should be a long barrier dam built across the entrance to the inland sea, across what was called the neck of the Zee; and once its inner waters had been tamed and rendered fresh and rid of much of its oceanic salt, portions of its southern, eastern, and western sides should be drained, and with the building of new dams and canals and the employment of massive pumps the like of which never seen anywhere before. Five brand-new polders could and should be created, on which cities could be constructed, and the nation vastly expanded and made safe from flooding forevermore.

Only a national government had the resources and the staying power to create and maintain such a massive project. It was to Cornelis Lely's imagination, engineering acumen, and eternal credit, and a mark of the Dutch government's prescience at well, that in 1917 his plan was formally accepted, and the nationally adored Queen Wilhelmina—then not even halfway through a reign which would last for fifty-eight years, longer than any other Dutch monarch—made the formal declaration from the throne: "the time has now come to undertake the damming off and reclamation of the Zuider Zee. . . ." Seldom has the mundane matter of drainage assumed such a heroic role in any nation's history.

The project started with the construction of an enormous barrier, twenty miles long and a hundred feet wide, built in a near die-straight line on a long northeast–southwest diagonal between the province of North Holland and that part of coastal Friesland that had been dried back in the twelfth century, and so already had some experience with the control of water. Damming the neck of the Zuider Zee meant effectively stopping up a highly active part of the North Sea itself—a notoriously badly behaved body of water racked with storms and swells and unpredictable deeps and shallows that has frustrated fishermen and navigators for centuries. The project—to build what was called the Afsluitdijk, the Barrier Dam—was very trying, time-consuming, and costly.

The project was officially begun in the summer of 1927. Some five hundred ships were pressed into service—an armada of iron dredges with chains and buckets and huge iron-ribbed hoses that tore up and sucked up from the seabed, deep beneath both the ocean and the Zuider Zee itself, immense loads of glacial till—boulders, cobbles, and gravel conveniently crushed during a previous Ice Age—and dumped them into two parallel lines in what was effectively the open sea.

The scale and Sisyphean difficulty of the operation is somewhat akin to trying to build a causeway across the English Channel—the tides and currents would wash away all before them as the fleets of dredges tried gamely to keep ahead of the sluicing powers of the water. Slowly, painfully slowly, the two lines of till rose under water and then, magically, they started to appear, fitfully at first, above the surface of the sea. Here they were pounded incessantly by waves, especially in winter. But if the barges worked tirelessly, so portions of the barrier were completed—first a few pinnacles, then some lengths, next long stretches of the barrier that eventually remained above the surface, coming to seem with time and effort, more or less permanent.

It was at this point that track-laying bulldozers clambered onto the still sodden tops of the twin lines of mud and laid between them the first of millions of tons of sand, taken from a flotilla of barges moored outside. They also tossed down into the water thousands of stone-weighted mattresses of willow switches—local Dutch farm families had a venerable tradition of cutting the long and flexible wands of *Salix viminalis* to make baskets—which could provide a secure foundation for the sand. In some unexpected places, however, there were problems with these osier mattresses—shipworms chewed them up, the iron bands that held them together rusted and broke—but in time these were repaired and restored and the till-protected sand dam began to rise higher and higher above the sea.

Lely's design called for the dam to rise to some twenty-five feet above what was called the Amsterdam Reference Level, to be certain of the fixity of what soon would be the *former* Zuider Zee, for if the project succeeded an entire sea would be permanently extinguished, its name to be erased from all maps and atlases and indexes. In its place would be a slowly freshening freshwater lake: its name would be the Ijsselmeer, after the river Ijssel, which flows into it from the east.

By the end of 1931, the Barrier Dam had just two gaps remaining—piercings that turned into foaming white-water chutes as the tides ebbed and flowed through them and scoured them deep. These last gaps were fearfully difficult for the engineers to close. The entire winter was devoted to trying to seal one of them by dumping a great thickness of all but impermeable

Ice Age boulder clay on top of the sand, then pouring mineral-rich soils on top of this, to be seeded in the springtime with thousands of tons of grass. The procedure worked. The sea's flow was indeed stopped. And one gap only then remained.

The triumphal culmination of the Barrier Dam project came on Saturday, May 28, 1932, when a flurry of barges moved around this last puncturing, and a forest of cranes and draglines worked feverishly in a maelstrom of wild water, and, under the pitiless gaze of dozens of film cameras, managed with a series of enormous splashes and slowly accumulating piles of boulders to staunch the torrent and finally seal its entryway, making it watertight.

At a stroke the Zuider Zee was formally extinguished, and forever. Cornelis Lely, creator of the dam, euthanizer of the Zuider Zee, birthfather of the Ijsselmeer, never lived to see his dream realized: he had died three years before, in 1929. But a statue of him would soon be thrown up at the west end of the barrier. He stands there today, proud and erect, booted and weatherproofed, staring down from a tower of basalt bricks as though willing the sea to stay put, daring it to breach his mighty construction. But at the same time, it is noteworthy that his ulster's hem is being blown back by a gale—whether for purpose of realism or as augury was never said. But it serves as a reminder of the forces of nature ever present in the Netherlands' story. And maybe a warning too, that this country above all is now standing into danger: that its very existence will be ever more tested as the world's sea level rises and the weather becomes ever more extreme, year upon year upon year.

Such thoughts were far from any Dutch minds back in the 1930s, of course. Back then, the North Sea outside might rage and thunder and ram its chest against the country—but so long as Lely's barrier dam held firm, all the innermost parts of the Netherlands would remain unscathed and untroubled. Moreover, with the dam completed, an array of new and habitable territories could now be created. Solid earth could be raised up from the depths and fashioned into something which to this degree and on this scale had never before been created, and by human agency alone, anywhere in the world: acre upon acre upon acre of spanking-new land.

The Dutch became formidably skilled at creating this new surface area, at manufacturing polders. For the Ijsselmeer they planned four: the relatively small reclamation known as the Wieringermeer tucked into the northwest of the lake; the Noordoostpolder, opposite it, at 120,000 acres, and Flevoland, split into two parts—Eastern Flevoland of 135,000 acres and its

Southern sibling, 107,000 acres. Work on Wieringer-meer was begun in 1927, before the North Sea barrier dam was fully finished. Work on the monumental pair of Flevoland polders in the south did not get under way until 1957—delayed in part because they were by far the largest and most ambitious of the new Ijsselmeer lands; but also because they would be vastly more expensive, and the postwar Netherlands had something of a trial to scrape together sufficient cash.

Once the money was in the Treasury's vaults, so the hard physical work began—and all of it, by the time the country got around to the making of Flevoland, conducted along well-honed lines.

First, long lines of twenty-foot-high dikes had to be built around the lozenge-shaped body of water that Cornelis Lely—now thirty years dead—had initially designed. The new island's point of origin would be in the southwest, close to the outer suburbs of Amsterdam, with enough free water between it and the port city to allow great cargo ships to pass in and out of the docks. The new island's northeastern end would be connected by a long bridge to the Noordoostpolder dredged dry some twenty years before, and which was now almost fully settled with farms and fields that stretched to every flat new horizon—horizons that exactly matched those of the sea which they had replaced, except that now, on occasion, there were trees, or in time the spire of a distant church.

It took some seven years to build the Flevoland dikes, and to ensure they were watertight. Along them were built an array of four widely separated pumping stations, named for the otherwise forgotten men—Messrs Lovink, Colijn, Wortman, and De Blocq van Kuffeler—who had long believed in the wisdom of the reclamation. In 1966 these four stations were fired up, and their nine gigantic mechanisms (four diesel, five electric) started sucking and siphoning the brackish water out of the swamp. One of the pumps was heroically lifting out some two thousand cubic meters of water—half a million U.S. gallons—every minute, night and day, and pumping it up fully eighteen feet into the surrounding sea.

In nine months, the polder was declared dry. The mud—thick and black and deep and deadly-quicksand-dangerous to anyone walking on it in its early days—was starting to solidify. The pools of remaining water were drying out in the spring sunshine. The mud was ready to become earth, then soil—and to that end Dutch government planes started to fly back and forth over the lifeless mudscape and broadcast thousands of tons of reed seeds, knowing that in the summer warmth of 1967 they would germinate and, in short order, cover the mud in a thick mat of greenery.

By the season's end thousands of acres were seen to be covered by chest-high stands of rushes, which when trampled down or, as later techniques advanced, deliberately burned down to create a rich mess of ash and stalks, evolved into area on which men could finally walk with safety. Without sinking. These first brave engineers thrust test rods down to see the depth of the gathering dryness—the reeds' transpiration would help mightily accelerate the drying process—and then devised and planned the routes of, and in turn excavated, drainage channels—for the pumps would continue running, as they still do to this day.

They sowed further clumps of seedlings, first of fast-growing willow trees, and then of moisture-loving pioneer plants like the celery-leaved crowfoot, the marsh fleawort, and the sea clubrush. Twenty-five thousand acres of Southern Flevoland were first covered with willow forests, which cunningly and speedily sprang up, a Dutch version of Burnham Wood come to Dunsinane. By 1968, eleven years after the first barriers had been hauled up, the entirety of the new polder was covered with thick vegetation—and with thickening soil that agricultural specialists now deemed ready to receive crops and to support those who might come to grow them.

But infrastructure had to be created. The dragline excavators and the bulldozers that came—at first half-sinking axle deep into the rippling expanses of mud, but soon finding their track-laying feet, as it were—would be followed by ploughs, seed-drilling machines, harvesters, heavy-duty tamping engines, road-rolling machines, asphalt layers, cement plants, curbstone makers, foundation diggers, architects, carpenters, town planners. Telegraph and utility poles would be erected, schools would be planned, as would libraries, hospitals, office blocks, endless rows of modest modern houses, and with farms built and widely separated on hundred-acre and two-hundred-acre plots. And there would be roads and bus routes and a long railway line, linking the new province with Amsterdam and the rest of Europe and the world.

And once all that was done, so the people were invited in, to the half-made landscape and the half-built towns. On January 1, 1986, Flevoland was finally formally open for business. Advertisements were placed in the newspapers in Amsterdam, Rotterdam, and The Hague, making it known that a new area of potential farmland was now available, and inviting applicants who wished to farm it.

And not just to farm it. But, and crucially, *to own it.* The state may have paid for all of this new land to be created, but it was not the state that ever intended to possess it. Such might well happen in the Soviet Union, or in China. It might well also happen in more nakedly capitalist societies, where choice parcels of state-made land would be handed out as party favors to the politically connected or to the ostentatiously generous. The Netherlands was signally different in its approach.

The Netherlands has for centuries been animated

Pancake-flat and perpetually windy, the twelfth province of the Netherlands, Flevoland, was raised from the sea and dried out in 1986, now supporting towns and cities, farms, railways and highways, and a population of 400,000.

by its so-called *poldermodel*. This is a uniquely Dutch philosophy, born out of the country's ceaseless need to do battle with the sea. It is a philosophy underpinned by a spirit of cooperation and compromise and a determined lack of evident privilege and separateness, and a concomitant belief that all of the population could and *should* share in such rewards as state-directed projects might ultimately generate. Popular communalism might well be the guiding ethos of the kingdom—but it would always be a communalism firmly yoked to individualism. New land would be owned not by the monarch or by the faceless official body that engineered its precipitation from the sea. Rather it would be owned by the best and the brightest and the hardest-working of the people who expressed their keen desire to own it. There was a distinctly Jeffersonian flavor to the *poldermodel*, so far as the distribution of territory was concerned—a belief within the Dutch psyche that it was important, as Thomas Jefferson had once written, "to provide by every possible means that as few as possible shall be without a little portion of land. The small land holders are the most precious part of a state."

Those who first came initially to live and farm as smallholding tenants on the new Flevoland plots would agree—and not at all reluctantly, it seems, for the Dutch are an obedient people by and large—to follow the guidelines of the government agency, the Ijsselmeer Polders Development Authority. For the first five years they had to plant and harvest scientifically decreed rotations of rapeseed, flax, peas, and grass—after which, if all worked out and the farmers made sufficient of a living and the polder soils proved to be as fertile as anticipated, the tenant farmers would be invited to bid for mortgages, for ownership, for the establishment of a vast omnium-gatherum of private property. After which, assuming the banks looked kindly on those who sought loans, the farmers would embark on the kind of intensive agriculture peculiar to the Netherlands, with nitrogen-fixing clover on the fallow fields, turnips for animal feed in winter, and then during the warm springs and summers, potatoes, beets, wheat, onions, and barley in abundance.

Some degree of social engineering was also attempted in the settlement of the polders. As recently as the 1980s, the Netherlands was divided religiously and politically by a venerable phenomenon known as "pillarization"—the Dutch word is *verzuiling*—by which the kingdom's three dominant groups, the Catholics, the Protestants, and the liberal social-democrats, were encouraged to develop along lines tinctured with a degree of what one might call cooperative segregation, of a mutually respectful separateness. This unenforced and somewhat vague sense of apartheid—one can see Dutch-influenced systems elsewhere in the world, South Africa and Northern Ireland more notably; and in the Netherlands it was until recently unthinkable for a Catholic and a Protestant to marry—was quite deliberately suggested to form a basis for the settlement of the Flevoland polder.

And so, the advertisements placed in the newspapers of the mid-1980s suggested, subtly rather than overtly, that it might be desirable and socially healthy for the ultimate demographic makeup of the region— and Flevoland would become the country's twelfth and final province—to reflect by its admixture the pillars of the nation as a whole. Those government officials who were charged with considering the thousands of applications for land took the religious and social affiliation of each family into account—with the result that the 400,000 who live in Flevoland today are a near perfect alloy of the nation's belief systems—one cannot say that Flevoland is Protestant, or Catholic, or liberal. Rather it is, quite simply, Dutch. It is, by its metrics, a near perfect distilled reflection of the kingdom as a whole.

And its capital—now a fully established city, thirty-seven minutes away by train from Amsterdam Central, and like so many new-made cities around the world dull, bland, flat, respectable, bustling, prosperous and inhabited by an evidently quite contented people—is the city of Lelystad. The city was named in honor of the man who, a full century before the water's floodgates were closed and the human floodgates opened, had this vision of making new land, in florid abundance, for the good of his people and for the security and stability of his country. *Si monument requiris*, one might say of Cornelis Lely as others had said of Christopher Wren in London, *circumspice*. If you wish to see his monuments—just look around you.

Except of course, what Cornelis Lely made was the very land on which everything else was to be constructed. Until his ambitious and revolutionary dreams of damming and draining and settling and farming took root and were publicly endorsed by a reigning queen, the making of land was an endeavor assumed to be exclusively in the purview of the Gods, or of Nature. Given the ephemeral nature of the Netherlands, coupled today with the ephemeral nature of all the world's land that happens to lie close to the ever rising sea, it may yet turn out that *lasting* is not necessarily going to be the singular feature of land that most appositely describes it. Land may turn out to be quite as temporary as the planet on which it lies. An irony that will not escape any Netherlander today, most especially if he chooses to walk across the Barrier Dam when a storm is brewing, and the North Sea starts to lick hungrily at his country.

3

Red Territory

To have and to houlde, possesse and enjoy and singuler the aforesaid continent, lands, territories, islands, hereditaments, and precincts, seas, waters, fishings, with all and manner their commodities, royalties, liberties, prehemynences, and profits that should from henceforth arise from thence, with all and singular their appurtenances, and every part and parcel thereof, unto the said Councell and their successors and assigns for ever.

—ROYAL CHARTER,
THE MASSACHUSETTS BAY COLONY (1628)

If a white man had land and someone should swindle
him, that man would try to get it back and you would
not blame him.

—STANDING BEAR, Ponca Indian leader,
address to the 1879 trial in Nebraska,
in which Indians were finally declared
to be human beings

S uch concerns as might trouble a modern Dutchman were not, however, on the minds of the hundreds of American men and women who in the early
spring of 1889 sat impatiently on their horses, and on
the driving seats of their wagons, and in squadrons of
waiting steam railway trains drawn up by freshly built
wooden platforms in the heart of the midwestern prairies. It was the April 22, a chilly Monday, a fateful day
in the history of the distribution of American land.

The crowds, some formed into ragged and ill-
disciplined lines, were all waiting for the coordinated
declaration—by mounted riflemen and buglers of the
U.S. Cavalry—of the hour of noon. At this precise and
presidentially ordered moment, when cannon were
detonated and signal shots were fired and flags were
dropped and brass instruments blown, the hundreds

would all be released, like racehorses at a steeplechase gate. And once let slip they would gallop, careen, speed, or chuff-chuff-chuff in a mad dash. There they would spring into action, each to lay claim later that day to the ownership of a piece of land, supposedly in their knowledge unpeopled, undeveloped, and unowned, and that each supposed would pass into their exclusive possession and remain so now and for always.

So, the horses would be champing, their riders would be checking the girths and the bridles, the train engineers would be readying themselves to swing open the regulator valves and turn the wafts of steam into billowing gouts of the stuff. Ahead of them all was an expanse of territory that would soon be a state named *Oklahoma*, a conflation of the Choctaw words meaning "Red People." The crowds—single men and whole families, drawn from all over the country and beyond— were engaging this April day in the first of many land runs, so called by the United States government, that had been designed to settle this territory's acres and, with hard labor and commercial guile and over hopefully brief periods of time, make them productive and pleasant, civilized and rich.

The great majority of these would-be settlers had chosen to believe what the government agents had told them: that the land ahead, and to which they would soon be laying claim, had previously been owned by no one. It was theirs for the asking, theirs for the taking. It was about to belong to anyone with the spirit of adventure, the pluck of the pioneer, who was equipped with a swift horse, a good eye, a sharp stick, and a large and unmistakable white claim flag. It would be virgin grassland, already officially surveyed and mapped and primed and just now quite ready for the plough and the sickle and the construction of a homestead. It was untouched and unclaimed, as government agents promised. It would be as unspoiled and unused as would be the willow forests of Flevoland a century later. What the hopeful were about to claim was brand-new land, entirely claim-free, with these waves of new settlers the first humans ever to inhabit it.

Nothing could be further from the truth. What transpired that April day in Oklahoma—arguably by numbers the most populous Native American state in the present-day United States—needs to be seen, as it seldom is, in a greater context. For in common with all the habitable quarters of America, the prairies in these parts had in fact long been well and truly settled, and to the extent that ownership—a concept eventually to be codified and regulated by the new nation's new government—was sympathetically understood by its native inhabitants, it was well and truly owned as well.

The North American continent to which white Europeans came in the sixteenth and seventeenth centuries had long been abundantly populated by a vast array of aboriginal native peoples. Today's estimates range from two to twenty million living in what is now the United States. They were by all current accounts well-organized into bands and tribes of considerable sophistication, agriculturalists in the main, farmers who lived in or beside small towns, and who had a profoundly deep attachment to the land from which they derived their various livings. When those who lived on the coasts first glimpsed with astonishment the white sails of the tiny inbound flotillas, they can have had no apprehension whatsoever of the savage interruption—of the apocalypse, indeed, the holocaust—that was about to despoil their serene and settled lives.

The very first white outsiders to arrive—by general agreement, the expeditions led by the Spanish conquistador Ponce de León, whose ships made landfall in 1513 on both the western and eastern shores of what he christened "Florida"—were however suspicious of those they first encountered, and in their behavior toward them were disdainful, intolerant, and short-tempered. The expedition deputies claimed to have found the local Indians—a people called the Calusa—markedly hostile, and there was much unseemly behavior—capture, killing, enslavement—in consequence.

The Spaniards may have formally claimed for Spain such of the territory as they saw, but they abandoned their first explorations of North America without creating a colony—this Ponce de León did with a second expedition eight years later, in 1521—and sailed off southwest in the hope of reaching Mexico, and there find compatriots. But they were shipwrecked off Galveston Island, in present-day Texas, where they met groups of local nomadic Indians, most probably members of the Karankawa band, who were rather more amiable than the Calusa, took pity on the starving sailors' plight, fed them fish and nuts, and offered entertainment and shelter. But in time the Spaniards turned on these people too and fighting and eradication once again dominated the relationship.

Neither the Karankawa nor the Calusa people exist today, evaporated and exterminated by imported illness and skirmishing. The Karankawa had been seasonal drifters; the Calusa, if they ever fully embraced the concept of landownership—which was somewhat unlikely, since they were a shoreline people, their homes perched on stilts among the mangroves, their habits more connected to fishing than to the growing of crops from the earth—it is a memory now long extinguished. What was once the exclusive domain of Floridian fisherfolk now gleams with the elaborate mansions and apartment buildings of the wealthy American retirees who have come south to cities like Tampa and Naples, to escape the northern cold.

The early Spaniards who first colonized these parts were in their behavior toward the locals something of a dishonorable exception among early European adventurers: it can be fairly be said that their conquistadores did not get matters colonial off to a good start. By contrast the early explorers and settlers from the more northerly European states—though they were quite as imperially motivated as the Spaniards, and so by today's standards equally deserving of contempt—generally liked what they found and were well liked in return.

"A gentle and loving people" remarked one of Henry Hudson's fellow Dutchmen in 1609, as his little ship the *Half Moon* wound its way upriver through the thickly settled estates of the Mohican people. And the Mohicans returned the favor: "Our forefathers immediately joined hands with the people in the vessel," a tribal elder told a conference a century and a half later, "and became friends."

Across on the far coast, too, there was an initial atmosphere of near collegial amity. In June 1579 Francis

Drake, privateer and explorer and Queen Elizabeth's then current hero, was beating his little carrack the *Golden Hind* northward through the eastern Pacific. He was hugging the coast, for reasons of both of navigational prudence and imperial possibility. He was well aware of the Spaniards' dominant Pacific presence south of the equator, as well as their being settled in their possessions—like the Philippines—on the far, western side of this immense body of water. But they had thus far made few significant inroads in the northeast—aside from Juan Rodríguez Cabrillo's sighting of the spectacularly useful bay at what is now San Diego—along the ocean's northern and eastern coasts. Drake wanted to lay claim to these untaken parts, and so sailed on northward (his crew one day noting with dismay a fall of snow, suggesting to modern historians that the ship went very much farther up the coast) and eventually brought his craft to a halt in what is now believed to have been northern California. The precise landing spot remains a matter of argument, but most probably it was somewhere between San Francisco and Mendocino, where a bay near Point Reyes, and now named for him, is edged with cliffs that bear a striking resemblance to the white chalk cliffs of Dover. Drake spent forty days here and dealt on terms of some intimacy with the local Indians—Coast Miwok, most likely—and professed himself entirely charmed by them. They were even more enthusiastically captivated in return, crowning Drake a king and making as to worship him and all his crew.

But there was a sting to Drake's apparent affection. He certainly described his hosts with sympathy, the men and women he met as being "of a tractable, free and loving nature, without guile or treachery." It turns out that he, being more rogue than saint, would behave with both treachery and with guile himself. He took formal possession of their lands in the name of Queen Elizabeth, managing to do so by claiming that the acres were a gift, and from a people who, so lacking in guile, did now "freely resign their right and title in the whole land unto her Majesties keeping."

"New Albion" was what Drake called his possession, and he had his ship's smith hammer out a brass plate formally stating his claim, with an English sixpence welded into it to give a semblance of Elizabeth's royal imprimatur. He wrote later that he had fixed the plate to a stout post somewhere near his campsite. But it has never been found—though a forgery of credible quality was uncovered in the 1930s, fooling many people until later science disproved its authenticity.

Drake's claim—a monstrous cheek, one might say today, since it is surely unthinkable that any

Miwok would seriously have agreed to handing over land on which their people had lived contentedly for generations—was to have profound implications for the future of Britain's relationship with North America for centuries to come. It was, to some historians and legal scholars, an indication that the whole country, east coast to west, had been taken by England and so the entire nation—most of its interior at this moment quite unexplored—was a colony.

Drake's taking of the Miwok land was of course not thought of as cheek at all back at court. In 1580 the grateful Virgin Queen, beneficiary in chief of Drake's American discovery, awarded her captain a knighthood. and touched her sword upon his shoulders in a ceremony on the *Golden Hind*'s afterdeck, in the very port of Deptford, in east London, from which the now *Sir* Francis had set sail four years before.

England's initial local intentions may well have been good. Those in the imperial vanguard may indeed have entertained publicly positive thoughts about such natives as they met, may have come initially to think of those on both American coasts, Mohican to Miwok, as gentle and innocent, as souls to be treated with respect and kindness. But this was the publicly expressed view. Deep within the newcomers' minds was a shrewd, cynical, and calculating component. For the Indians were their rivals, competitors for the possession of the great tracts of the square mileage on which they had for so long existed, but which the incoming settlers now so dearly wished for themselves. And as early as the sixteenth century the newcomers began the lengthy and initially stealthy process of wresting away that land.

It was a complicated and accelerating process that ranged from prudent exchange by way of cunning purchase to fits of outright theft. It came mostly veiled in a pervasive attitude of curt dismissiveness, which, in essence, would define almost all imperial adventuring among those the newcomers thought of as inferior. It led to a series of processes—followed both by the colonists and then in ever more dramatic fashion by the new-made Americans themselves as they ambitiously pursued their Manifest Destiny—that would eventually deprive an ancient native people of almost all of the territory which had given them such sustenance for many thousands of years.

Those who led the first two fully organized and lasting English settlements on American soil—the Virginia Company, with its original headquarters at Jamestown, and the Massachusetts Bay Colony, first based in the seaport of Salem—enjoyed a growing certainty about their right and duty to gather, claim, and own—in the name of the English monarch, that is—the land

that they would now administer. The Dutch human-
ist scholar-poet named Hugo Grotius had advanced
a legal opinion that offered broad justification for the
settlers to plunder as they wished: "land which lies
common, and hath never beene replenished or subdued," he wrote in a text considered in Europe of the
time to be the essence of wisdom, "is free to any that
possesse and improve it."

Not a few of the Puritans were, however, somewhat
discomfited by the employment of this blanket argument of consent. so they sought additional biblical authority. They had considerable early admiration for the
local Wampanoag people and had an understanding
that these friendly and cultivated natives—who were
decidedly helpful to the nervous strangers now in their
midst—had a particular reverence for the land that
they had clearly been using for a very long time, and
with evident sophistication for agriculture. Massasoit,
the great sachem of the tribe, described their feeling
toward the earth beneath their feet. The land, he
declared, was "our mother, nourishing all her children,
beasts, birds, fish and all men. The woods, the streams,
everything on it belongs to everybody and is for the use
of all. How can one man say it belongs only to him?"

John Winthrop, the Bay Colony's second and most
influential governor, was the first to dilute Grotius's

opinion by stirring a tincture of godly authority for landownership* into the mix by arguing—as the English philosopher John Locke would argue more forcefully fifty years later—that it was man's Christian duty not just to own the land, but to improve it. The Book of Genesis, after all, spoke of the injunction for humankind to "increase and multiply, replenish the earth and subdue it." Winthrop could thus argue that the settlers now had both a natural, God-given right to own, but a civil right too, one that came about only when the land was *fenced in, manured and improved*, to paraphrase his own words in a pamphlet that he published in 1629

---

* The traditional views of the Wampanoag toward landownership echo many ancient attitudes to land elsewhere, a notion that will be explored later. Jews, for instance, drew their original inspiration from the Book of Leviticus, and Yahweh's commandment that all land was God's and that "strangers and sojourners" had no innate right to possess it. Then again, Hammurabi ruled Babylon by courtesy of the Levantine deity Baal, and any citizen wanting land nearby could acquire the tenancy of it only by serving the king in his role as god's agent. Likewise, in imperial China, the Celestials won their use of territory in return for service to the Son of Heaven, who owned the country on behalf of God. And even in the less devout corners of Europe the barons and lordly bishops saw themselves as in service to the Almighty, and those whom they permitted to occupy land were no more than components of a long string of obligation connecting Heaven to Earth, by way of Man.

designed for the westbound voyagers to read on their Atlantic crossing.

Captain John Smith—the man who coined the phrase *New England* for the territories to the north and west of the Plymouth Colony—was even more forceful than Winthrop about the pioneers' need to acquire land, and he had little truck with the idea that the Deity should play any kind of role in justifying its seizure. His experience in Virginia—where he had understandably conflicted feelings about the local natives, the Powhatan, since they had been planning to execute him, and would have done but for the legendary supposed intervention of the chief's daughter, Pocahontas—led him to believe that the English should be allowed to live more or less where they liked and to own what they would. America had an abundance of room for all, he told his followers, and so far as he was aware this room was quite unoccupied. If this single fact wasn't sufficient to ease their consciences, they should remember, he cautioned, what little monetary value the Powhatan, the Wampanoag, or any other Indians so far encountered had placed on it: "for a copper knife and a few toys, as beads or hatchets, they will sell you a whole country—and for a small matter their houses and the ground they dwell upon."

And there was one other justification for land seizure, and that could be added to those permissions offered by international law, by the Bible, and by simple need. There was, for the English most of all, the matter of the king, and the "divine grace" by which he reigned and ruled. The formal royal charters that set up the colonists' adventuring in the first place gave the settlers the right—and indeed the duty—to take such land as they wanted in the king's name. They would create real estate—royal estate, some might say—in the unquestioned name of God.

Superimposed on the specific ideas that then justified native land acquisition, and which were advanced by colonial managers like Winthrop and Smith, was a very much more ancient legal provision of breathtaking audacity, known generally today as the Doctrine of Discovery. This notion was first adduced by the fifteenth-century Portuguese, who sought and won a papal bull—known as *Romanus Pontifex**—which permitted

---

* This bull issued in 1454 by Pope Nicholas V—confusingly one of two that he titled "the Roman Pope," the other being quite unrelated and dealing with the question of Jews living in Austria—allowed and indeed encouraged the Portuguese to take any land they discovered in Africa south of Cape Bojador (the significance of this famous maritime landmark, a cliff jutting out from Western Sahara, has been explained in my book *Atlantic*). The bull also sanctioned slavery, though only if the peoples enslaved were non-Christian.

them, very basically, to seize as much of west Africa as they cared to take. The idea put forward then—and which has generally held legal sway throughout all subsequent American history, even though it started by being limited to the colonizing ambitions of just Portugal and Spain—held that if a European nation was lucky enough to discover foreign lands peopled by non-Christian natives, it had an inalienable right to own those lands and to colonize such people as were unlucky enough to live in the Europeans' path. England and France soon followed suit, employing their own version of the Doctrine of Discovery to justify their own behaviors around the world as their empires burgeoned through the centuries; and the United States, once it got itself under way in the 1780s, used the same or a similar version of the doctrine to claim ownership of the lands hitherto settled by Indians. Merely by planting a flag, the doctrine said, you owned the land on which you planted it.

The British went further: believing the doctrine extended authority over humanity every bit as much as it did over landscape, they also took large numbers of Native Americans into slavery. Ironically the Spaniards, reputedly leaders of thought in these matters, had made slavery of Native Americans illegal in the mid-sixteenth century, leading the British to employ subterfuge and semantic sleight of hand to force indigenous people to work their plantations in the Caribbean—fulfilling the initial land-grabbing and slave-making suppositions of the doctrine to the letter, in other words. The Americans later took thousands of locals as slaves too, though this "other slavery," as it has been recently called, was also perpetrated in cunning ways, coercive labor based on financial indebtedness being the classic means of forcing natives to work on the southern plantations, alongside the African importees.

So now the principle had been conceived and agreed to—that by divinely sanctioned right and with formal papal endorsement, European flag-planting settlers might now take away the land previously settled by the America's vast and varied population of native peoples. And with that, so the rush to gather up the spoils—the land and its landscapes, soon to be deconsecrated and commodified—got ponderously under way.

Land was by now a fully recognized capital asset, with its value positively straining to be released by the labor of men who worked on it. This was one of the central messages of the then widely published and popular writings of William Petty, the seventeenth-century philosopher-anatomist who was at the time cutting a swathe through economic thinking in England. Petty—mathematician, linguist, astronomer, surveyor, navigator, doctor, thinker of formidable breadth, and a prodigious talent*—had convinced most in the English establishment of the day that land was not a merely static resource, rock and grass and little else, but was an asset with an inherent capital value that could be released by the simple process of working it. Moreover, land's intrinsic value would also increase by the simplest principle of supply and demand, in that land was limited in supply but the population that wanted it would always be guaranteed to increase, such that the land would become ever more in demand, and so command higher and higher prices at sale. Since land would in both theory and practice bring profit to all involved, land must be acquired by colonists coming to America

---

* William Petty's early fame arose from a curious incident in 1650 when he was professor of anatomy at Oxford, and which involved his proposed dissection of a young woman, Ann Green, who had been hanged for alleged infanticide—her weight on the public gibbet augmented by her friends clinging to her legs for fifteen minutes to hasten her death. As Petty prepared his scalpel, he heard a rattle in her throat, realized she was still alive, gave her brandy, massaged her legs, for some curious reason gave her an enema and brought her back to full consciousness. She lived for twenty more years, married and bore three children—always professing her innocence of murder and claiming her infant had been stillborn. Most—given the miracle of her own revival—were inclined to believe her, including her savior.

if they were to profit and prosper and lead the colonies into the sunlit uplands of economic success.

The colonial managers agreed entirely. For a while, any settler who paid his own way across the ocean to Virginia was given a so-called headright of one hundred acres of land, free and clear. So popular did this enticement become, and the ships in consequence so crammed to the bulwarks with migrants, that Virginia had in short order to cut back on its promise and instead charge a modest five shillings for fifty acres—and even then most of those flocking to the New World were said by the historian Robert Beverley to be "not minding anything but to be masters of great tracts of land."

A lust for land was from now on part of the quintessential allure of America, a lust that transcended the more principled reason for migration, which had been the religious and political freedoms promised by breaking free of the stiffening constraints of the older world. Land was what could be had, and in abundance. Except, of course, for the inconvenient reality that this land was already owned by others, by communities who had a prescribed right to have and to hold on to it.

So, what would eventually happen in Oklahoma, and which was centered most profoundly around the distribution of land, was now firmly on the horizon.

What would be particularly egregious examples of land grabbing now simply required the passage of time—a century and a half of further colonial rule, then the creation of the United States out of the divided remains of the gigantic American fiefdom of the British empire, and a century more of Washington's independent governance, briefly fractured though it was during Abraham Lincoln's presidency. During all of these periods of North American history, native lands were being thinned and culled and native people were being slain and pinioned into remote reservations, tracts that were often waterless or infertile or both, impenetrable laagers that no white man would ever wish for. Or else they were forced away to become homeless, or were subjected to savage legislative sanction—leaving the country in the state it is today, with (in the view of many critical environmentalists) the original inhabitants left mainly landless, and those who have since occupied these former native territories disrespecting and despoiling them in a manner no Indian ever would have done.

For the Native Americans had looked after their land. They tilled and fertilized it, they grew ample harvests of corn and squash, beans and cotton. They dug irrigation canals. They used fire—most bands had flint carriers who could set fires to trap and encircle animals, to clear forests for planting, to illuminate the night. They built roadways and trading networks, and even the forests were culled to space out the trees so that bands of men in large numbers could move through them at speed. Their villages had accommodations both permanent for the settled tribes and portable for those still practicing nomadism. They had tribal governments. They held ceremonies and councils and meted out justice and maintained discipline. They worshipped gods, built temples, carved totems.

They were, in short, a sophisticated and civilized people—and though many Americans today believe there to be precious few natives remaining in the country, there are in fact some five hundred tribes remaining and officially recognized today, with as many languages spoken, and of as rich and complex a set of etymologies as might be imagined. Native Americans are all, it should be remembered, genetically related to those who twenty thousand years ago crossed by way of Beringia, the land bridge across the north Pacific between eastern Russia and western Alaska. Native Americans are, genetically, a Pacific people. They are, in essence, an Asian people. Americans are Asians—or they were, until the Europeans swept in and stole so much from them, and came to regard the Indians essentially as savages, as barely human primitives in the country that they had first settled, the country that they had first made.

The subsequent and now fully permitted trade in their land—with private European individuals persuading Native Americans to part with their holdings, by way of a hastily written treaty or bill of sale, and invariably for a pittance—went on apace for well over a century after the founding of the first American colonies. However, in the critical year of 1763, this all came to a screeching halt—or at least, it was briefly circumscribed and regulated—and by the involvement of no less a figure than England's eccentric but kindly farmer king, George III.

He issued a formal proclamation from London, which very basically declared that no English settlers could henceforth seize, buy, or settle any lands that lay to the west of the ridge of the Appalachian Mountains. The already thickly settled land of the thirteen colonies between the mountains and the sea were deemed sufficient for the time being. All to the west—where the colonists, in the royal view, had no serious and sustained interest—would now be reserved for the Indians alone, and would be lands where they could henceforth live untroubled and undisturbed.

The Proclamation Line, as it was called—porous in the extreme, and in short order widely ignored—

served precious little purpose, even if it briefly stilled some of the more egregious attempts at westward expansion. Quite to the contrary: many consider it to have been one of the many factors that so infuriated the settlers—for how dare London dictate where in the Americas they might live! George Washington was one such who was angered by the proclamation: as a professional surveyor he could well recognize the finer and more valuable tracts of land he encountered, and when bounty from his participation in the French and Indian War came in the form of land, he gathered some choice tracts for himself—a total of 32,000 acres of fine farmland that was now most inconveniently sited to the west of the Proclamation Line, and which he was now ineligible to possess. The consequences of his irritation at such a colonial ruling were legion: the 1776 Declaration of Independence and the six years of fighting between redcoats and patriots that followed have the matter of Indian lands as one of its many causal origins, and George Washington's eventual leadership—and presidency of the new republic—a natural culmination.

The Proclamation Line was drawn at the end of Britain's victorious Seven Years War with France—or the French and Indian War, as it was named in North America, since various Indian tribes allied themselves with the two warring sides.* When Britain did win all of the hitherto French-held territory, this so-called Ohio Country—the huge and richly fertile landscape much admired by Washington, dominated by the Ohio River at its center, lying between the Appalachians and the Mississippi River—was transferred by treaty to British rule. Such Indians as lived there were understandably perplexed by the sudden change of their rulers' identity, angered by being passed like chattels from one empire to another. One day a fort from which French soldiers might emerge to make nuisance had its national colors struck down and replaced with the Union flag of Great Britain, and out of it would stream murderous redcoats. It was all most bewildering.

The British soon came to be widely regarded by the Indians as much less congenially disposed than the departing French. Infamously, Jeffery Amherst, the British commander in chief during the closing stages of the campaign, had little time for the Indians, regarding them as "more allied to the Brute than to the Human Creation." He happily agreed with one of his colonels

---

* Among the Indian tribes aligning themselves with the French were the Algonquin, the Ojibwe, the Lenape, Shawnee, and the Mohawk. The British, meanwhile, benefited from the support of the Iroquois, the Catawba, and the Cherokee. But there was much volatility and side switching.

to supply the Indians besieging Fort Pitt—later to be Pittsburgh—with blankets soused with smallpox bacilli. "You will Do well to try to Innoculate the Indians by means of Blanketts,* as well as to try Every other method that can serve to Extirpate this Execrable Race."

It was with many of its leaders endowed with attitudes much like those of Lord Amherst that, a decade and a half later, the newly independent United States of America took on the challenge of dealing with its original native inhabitants. It was to be a deeply troubling story. The Indians had fared badly enough under the British; they would face more than two centuries of near unrelieved misery at the hands of the vigorously expansionist Americans. For while the British were generally—or at least, by comparison with what was to come—somewhat geographically timid in their westward venturing, the new Americans unleashed themselves on the territory with improvident glee. It was after all their Manifest Destiny—the stir-

---

* Amherst College, named for the town of Amherst, Massachusetts, in which it is sited, voted in 2016 to end all public association with the now disgraced English soldier, and to abandon use of its sports team mascot "Jeff." Controversy naturally followed the decision, with traditionalists decrying the proposed change.

ring phrase was coined fifty years after the United States was born, but it was sentiment keenly *felt* almost from the get-go—to sweep the civilizing light of the new nation through every dark corner of the continent.

Settlers and speculators alike looked hungrily westward. First in their sights were the fertile lowlands of the Ohio Country, the land lying between the Appalachians and the Mississippi River that had recently been won from the French. Next, following Thomas Jefferson's celebrated Louisiana Purchase of 1803, it was the enormous tract of land on the far, right bank of the Mississippi, which extended all the way to the Rocky Mountains, and which, though little explored and surveyed, should surely provide rich pickings for anyone foolhardy or brave enough to venture out to make his fortune. Then there were the Spanish territories of Texas and California—millions upon millions of acres of land that looked at first quite freely available to any white man who wished to settle and prosper. If any native peoples primitive or reactionary or inexplicably unwilling to be annexed and subjugated happened to stand in the way of this God-given right and duty, then they should, said the settlers and the speculators, be brusquely swept aside, all in the name of progress and the common good.

The Founding Fathers had made one crucial provision in the very first article of the U.S. Constitution—in Section 8 lies the sole reference in the entire document to the matter of Native Americans: "Congress shall have Power . . . to regulate Commerce with foreign Nations and among the several States, and with the Indian Tribes." The rule of thumb in American law holds generally that matters not specifically reserved are for the various American states to regulate and legislate. Had "Commerce . . . with the Indian Tribes" not appeared in the Constitution, it would have been up to the state of New York, say, to deal with its resident Mohawks, the state of Florida to do business with its Seminoles, South Dakota to try to reach settlements with its various tribes of the Sioux, and so on. But because the words do appear in the Constitution, and quite specifically, so the question of dealing with the various tribes has been, from 1790 onward to today, a matter for the federal government alone. So the business of dealing with the 66 million acres of land currently set aside for Native Americans—a somewhat shockingly tiny 2 percent of the entire landmass of the United States (shocking because the *entire* landmass once "belonged" to its native peoples, in theory and by natural right)—is for the politicians and bureaucrats in Washington, D.C., to decide.

And for the first century or so of the United States'

existence, until the beginning of the twentieth century, the manner in which Native Americans and their land-holdings and land claims were dealt with was decided essentially by the three elements of Law, Treaty, and War. In all three categories, the white man won and the Indian lost.

A complex web of federal laws passed during the nineteenth century asserted and encouraged the rights of Americans to own the land that, since the Louisiana Purchase, now stretched near limitless, over to the western horizon. These ownership laws applied, notably and with the cruelest of ironies, to American citizens or those in the process of applying to be American citizens—but not to Native Americans, who in the main were still technically *not citizens of their own country.* Those few who by quirk or exception owned individual tracts of land, those who married outside their tribe, who abandoned their tribe, or who were of mixed blood, could in theory acquire citizenship. But it was not until 1924, and passage of the Snyder Act, that the three hundred thousand Native Americans then living in the United States could become, automatically and without application, full citizens. There can be few today who would consider this state of affairs anything other than shameful.

Similarly repellent, and in the specific matter of land

rather than of citizenship, is the U.S. Supreme Court ruling of 1823, which is taught in the first-year classes of most law schools to this day, and is known familiarly as *Johnson v. M'Intosh*. The case itself is quite complex, its facts muddled by some fast-and-loose chicanery by one of the parties involved. But the court's unanimous decision, announced by John Marshall,* the hugely distinguished chief justice, held in essence that only the federal government may purchase land from Native Americans. Indians may not sell land to private individuals because, though they themselves are allowed to live on their traditional lands and enjoy "aboriginal title" to them, that title is precarious and extinguishable only by the United States government, and then only by "purchase or conquest." The Doctrine of Discovery, as noted above, held that the government was the only truly legitimate ultimate owner of the Indians' land—and this 1823 decision formally ratified and confirmed this arcane belief. The Indians held title, but not quite the same kind of title—saleable, transferable, alienable—that white people were allowed.

---

* Marshall, a Virginian who had been one of the country's Founding Fathers, served for thirty-four years, from 1801 until 1835, the longest term for a Chief Justice to date. He was appointed by John Adams, the country's second president, for whom he had also served as secretary of state.

It was through later legislation like the Distributive Preemption Act of 1841 that white settlers were able to ease their way into ownership, and the United States to expand its populated self ever westward. This law gave half a million acres of federal land to each of nine designated midwestern states. It then allowed any citizen who was squatting on that land the right to buy it from the government for $1.25 an acre, so long as he pledged to work it and improve it for a minimum of five years (if he didn't, the federal government could take the land back). Ten percent of the total revenue from any sale was then handed to the states to help build roads, bridges, canals, and railroads. It was a system that proved of enormous benefit to the idea of the Manifest Destiny and it did considerable good to all who took advantage—but not, of course, to those who were specifically excluded from ownership.

Twenty years later came the Homestead Act, signed into law by President Lincoln in 1862. It was designed to ease the process of ownership, allowing those who never had squatted on the land—people living miles away in cities, for example—to acquire quarter-section lots of 160 acres, for no more than a tiny registration fee with the United States Land Office. Freed slaves could apply; Indians could not.

And the same was true when it came to constructing the great and very costly railroads across the prairies. The government gave grants of land—whether it was theirs to give still questionable to those who were there first, of course—to the railroad companies, immense acreages extending along hundreds of miles in ten-mile-wide strips both sides of the proposed lines of the permanent way, and which the companies would then use or dispose of as they wished. Almost all sold the land as they needed, raising millions in cash from settlers well aware that holding land beside a railway line could possibly turn to their commercial advantage, eventually.

Company schemes—done honestly, by and large—fully intended to separate would-be settlers from their money were legion, and most seductive. The Burlington & Missouri River Railroad Company, for one, published handsome broadsides with full-color engravings of prairie scenery, offering "Millions of Acres" of "cheap land," in the flatlands of Iowa and Nebraska, and on most tempting terms: less than $2.50 an acre, with ten years credit at 6 percent interest, a seventh of the principal price due each year, but payable only beginning four years after the purchase. Moreover, "Land Exploring Tickets" were sold at stations, the price refunded to anyone who bought land as a result of what they saw. And anyone flush enough to pay in cash would get a fifth off the price. The Burlington Railroad first published these broadsides in Buffalo, New York, far from Nebraska, hoping to attract immigrants from the crowded east to help settle the near empty midwest.

But it was only white men and women who could make such purchases and come to own the land. Ownership, like voting, was then for citizens alone, and would remain so for decades more. Others would have to wait. Indians were still on the outside: the yeoman farmers of whom Thomas Jefferson had wanted his country made had to be of moderate respectability and racial acceptability, and most certainly not of aboriginal stock.

Much the same lack of fairness applied to the bewildering matter of treaties, the solemn, red-morocco-bound, gold-blocked, and sealing-wax-adorned documents that reside still in the safety of the National Archives, pregnant with permanence and supposed meaning. The great majority of the 368 treaties that were signed between the United States and the various tribal nations—a practice that the United States ended in 1871, with many of the treaties either subsequently broken or else never ratified by the Senate, so unenforceable—involved land. A classic example of how the process worked was related by Philip Deloria, a Harvard professor and member of the Yankton Sioux

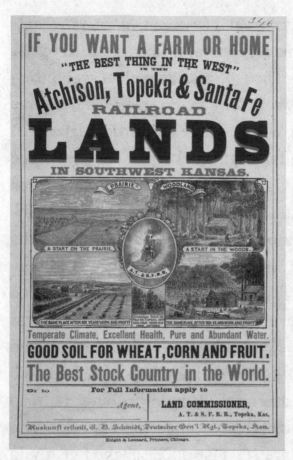

IF YOU WANT A FARM OR HOME
"THE BEST THING IN THE WEST" IS THE
Atchison, Topeka & Santa Fe RAILROAD
LANDS
IN SOUTHWEST KANSAS.

PRAIRIE   WOODLAND
A START ON THE PRAIRIE.   A START IN THE WOODS.
A. T. & S. F. R.R.
THE SAME PLACE AFTER SIX YEARS WORK AND PROFIT   THE SAME PLACE AFTER TEN YEARS WORK AND PROFIT

Temperate Climate, Excellent Health, Pure and Abundant Water.
GOOD SOIL FOR WHEAT, CORN AND FRUIT.
The Best Stock Country in the World.

Or to   For Full Information apply to
_____, Agent,   LAND COMMISSIONER,
A. T. & S. F. R. R., Topeka, Kas.
Auskunft ertheilt, G. B. Schmidt, Deutscher Gen'l Agt., Topeka, Kan.
Knight & Leonard, Printers, Chicago.

Keen to promote settlement in the prairies—and so create cities and a population of potential revenue-paying passengers—America's railway companies eagerly advertised the benefits of landownership in the wide open spaces of the west.

tribe from South Dakota—a tribe that signed away rights to their vast prairie landholdings in the southeast of the state in June 1858.

South Dakota is a state alive with Indian memories— most particularly, memories of the Sioux. These were Plains Indians, not homesteaders and farmers, not designated as civilized by the government, but instead feared for being fearless and splendid—tepee dwellers, ever moving, magnificent horsemen, masters of animals that the colonizing Spaniards had imported by the sixteenth century. Their best-known chief was Sitting Bull, his best-known victory that over General George Armstrong Custer at the Battle of Little Bighorn in 1876. And later, in the frigid cold of the last days of the century, scores of Oglala Sioux were machine-gunned to death at the Battle of Wounded Knee. The Sioux Indians had good reason to loathe the white settlers who came for the Black Hills gold and the chance to scourge the land of bison. They had ample cause to be appalled by the settlers' reverence for such violent figures as Calamity Jane and Wild Bill Hickok. They surely had some right to feel aggrieved that white men employed gangs of Chinese and Irish workers to build railroad tracks arrow-straight across their endless, hitherto peaceful miles of grasses that once soughed gently in the prairie wind. They came quite understandably to fear and despise such settlers and fortune-hunters, as well as the trigger-happy soldiers who were there to protect them but who all too often shot first rather than talked or ever tried to parlay. The hatred began, and with reason, back when the first lands were taken away, taken by cunning, deceit, and guile, at the events of 1858 that Philip Deloria recounts.

It had taken three months of hard negotiation, conducted in Washington, D.C., between government agents and seventeen bewildered Sioux leaders—no fools, though the officials initially supposed them to be—before an outline treaty was agreed to. This document, infamous on its face, held that fully eleven and a half million acres of Sioux land—grassland, rich with animals and birds and unknown and untapped minerals no doubt, as surveyors would later confirm—would be freed from Sioux control and opened for settlers to use as they pleased. Just four hundred thousand acres, the equivalent of a slumlord's studio apartment to a people who had hitherto inhabited a country estate—would be held back for the exclusive use of the Indians.

As all treaties did, this one, wrote Professor Deloria,

crunches them down geographically into a smaller space, opens up all the surrounding land to white settlement, and then sets up the provisions for a colonial kind of structure and the reservations. So, this is a reservation treaty. It creates a reservation, and it sets up all the structures that are going to go with the reservation.

This settlement is for $1.6 million over fifty

years. It's got a kind of sliding scale: $50,000 a year in the first ten years, and then a lesser amount, and then a lesser amount. But what happens in this Treaty, as happens in many Treaties, is a whole series of qualifications about the money, about what happens with that money. If you don't do this, we will reduce the money. If we decide that you need an additional agent or an additional farmer, we're going to take it out of your money. If you don't send all children between seven and eighteen years of age to school to learn English, we're going to take some of your money away. If you drink alcohol, we're going to take your money away. So, the Treaty goes on for many pages, laying out here's what you give up, here's the compensation for it. Oh, and by the way, we're going to fully manage the compensation; oh, and by the way, if you don't behave in all these very specific ways, we're not going to give you the compensation at all, anyway. So, it is very sad to read this Treaty, and read it thoroughly, because the fundamental piece of it is the land. And the next piece of it, the majority of it, is about the money and what's going to happen with the money. Then it's about altering Native behaviors, in absolutely essential ways.

Philip Deloria's great-grandfather was one of the Sioux signatories* of the treaty, which was signed in 1858 by the U.S. commissioner for Indians in what was then Dakota Territory, in the presence of ten white citizen-witnesses. The Senate ratified the document nine months later, and President James Buchanan added it into the list of formal treaties to which in theory his nation remains perpetually bound. The reservation today, six thousand generally impoverished people, forlornly advertises itself as "the Land of the Friendly People of the Seven Council Fires of 1858." Its busiest business is the Fort Randall casino in the city of Pickstown. Beyond the reservation borders stand millions of acres of rangeland, specked with cattle, almost all of them owned by white farmers. The Sioux can only now gaze outward, wistfully. At their home that once was.

At least the Sioux were left more or less where they were—albeit much diminished, much cheated, ill

---

* The Sioux signers all inscribed an X as their mark. Some of their vernacular names were lengthy indeed: TA-CHUNR-PEE-MUZZA, for example, or CHE-TAN-A-KOO-A-MO-NEE, translated obligingly as His Iron War Club and Little Crow respectively. Professor Deloria's great-grandfather was either Hairy Thigh or The Owl Man, though family memory is said to be a little vague.

served. But at least they remained in the prairies, where their ancestors had lived and hunted for generations. Many other Indian tribes were not so fortunate, suffering from the cruel and haphazard policies generally known as Indian removal—whereby Native Americans, regarded by whites as greatly inconvenient to their own ambitions, were requested, demanded, and then ordered to move, away, out of sight, and like dust, to be swept under the rug. All too few tribes today live on the lands they originally settled: the Mohicans of eastern New York, for example, have since 1824 lived a thousand miles away on a reservation in Wisconsin; the Ojibwe of Michigan were forced by government agents to walk hundreds of miles to a new site in Minnesota, where they were cheated of their promised food and money and walked back through the woods in the dead of winter, five hundred of them starving to death. And most infamously of all: in the late 1830s a hundred thousand eastern woodland Indians—members of the tribes of the Cherokee, Choctaw, Creek, Chickasaw, and Seminole—were brutally compelled to walk at gunpoint across the mountains and plains to settle anew in the unfamiliar expanses of what is now Oklahoma. Their cruelly coerced westward passage is infamous today as the Trail of Tears; there were many such trails, much less known, all bent on separating native peoples from their traditional lands, all indelible blots on the American escutcheon.

And westward was invariably the direction of their journeying. For this was the direction in which white men moved to fulfil their promised destiny. Westward to the ever shifting frontier, with the Indians moved ahead and into the unknown, beyond their own Pales of Settlement, and to places where, in the white men's eyes, they could do no harm except to their savage and miserable selves.

There were many architects of the removal plan. Thomas Jefferson, most particularly, and, in 1803, very early on in the American story, proposed the idea of offering tracts of government land to the Cherokee—a tribe then living in the west of the Carolinas and Georgia as well as in Tennessee—if only they would shift themselves west of the Mississippi. Each family would get a full section—640 acres of virgin territory—if they would agree to move. Other presidents had similar ideas—James Monroe and John Quincy Adams offered incentives for the Indian to move voluntarily. But then, and most notoriously, came the seventh president, Andrew Jackson, a Democrat who would have no further truck with an evidently recalcitrant aggregation of Indian feeling in the fertile settler country of the American southeast. He wanted all to go—particularly those acknowledged to be advanced, settled, and self-governing, and condescendingly known as the Five Civilized Tribes. These were the aforesaid Cherokee, the Choctaw, the Chickasaw, the Creek, and the Seminole. They were told—or at first politely asked, since the removals were theoretically voluntary—to head out, to pack up their belongings, to go along through the woodland roads they had built themselves and head for freedom, as they supposed, in what was designated officially by the government as Indian Territory.

President Jackson—derided in more modern times as the "exterminator of the Indians," and considered possibly even guilty of a genocide—tried to adopt a public tone of kindly persuasion, though read today his famous address to the Creek leaders seems oleaginous at best, and ponderously shifty:

Friends and Brothers—By permission of the Great Spirit above, and the voice of the people, I have been made President of the United States, and now speak to you as your Father and friend and request you to listen. Your warriors have known me long. You know I love my white and red children, and always speak with a straight, and not with a forked tongue; that I have always told you the truth. . . . Where you now are, you and my white children are too near to each other to live in harmony and peace. Your game is destroyed, and many of your people will not work and till the earth. Beyond the great River Mississippi, where a part of your nation has gone, your Father has provided a country large enough for all of you, and he advises you to remove to it. There your white brothers will not trouble you; they will have no claim to the land, and you can live upon it you and all your children, as long as the grass grows or the water runs, in peace and plenty. It will be yours forever. For the improvements in the country where you now live, and for all the stock which you cannot take with you, your Father will pay you a fair price. . . .

In their thousands, beginning in 1830 when twenty thousand Choctaw* set out from their home villages in Alabama, the walkers and shufflers, staggerers and swaggerers, striders and limpers set out on the trail. They took weeks, months, to complete the journeys,

---

* The Choctaw were accompanied by five hundred Black slaves. Disagreeable though it may be to recount, the Five "Civilized" Tribes were regarded as being advanced in part because they each owned slaves, thereby mimicking the behavior of the supposedly advanced whites of the time. The Cherokee took some two thousand slaves on their various marches.

and thousands died en route. Soldiers escorted them, along with government agents from the newly established Bureau of Indian Affairs, mainly to ensure none escaped, backtracked, or tried to sidle off and settle on such tempting pastures as they might glimpse while passing by, and which they would naturally be forbidden to enjoy.

Diaries of the escort soldiers tell of the collective misery. One, a Lieutenant Jefferson van Horne, writes of having loaded scores of Indians "and negroes" onto a boat to cross the Mississippi in December 1832.* On the December 4, after noting, laconically, "one birth," he writes that his party of 634 souls had covered twelve miles before camping. Next morning

> all the captains called on me in a body and desired me to wait until the cart of their head man Etotahoma (which broke down last evening and was unable to get to camp) should be brought up. I had

---

* The boat chosen for the Choctaw ferry was the *Heliopolis*, one of a pair designed by Henry Shreve, inventor of the iconic Mississippi stern-wheeler. Together with its twin the *Archimedes*, the craft was put to clearing away miles of "snags," thousands of hopelessly entangled tree trunks and branches that clogged the river, halting all navigation. Shreveport, Louisiana, is named for this great Quaker engineer and shipmaster.

sent back more than once and had much trouble to get this old man and his cart along. His oxen were poor and worn out, and his cart badly constructed. But he was looked to and beloved by the whole party. He would not part with his cart, and though it might have been policy to go on and leave the wretched old establishment, I found it impossible to get his people along without him. He was old, lame and captious and gave me more trouble than all the rest of the party. Proceeded at half past nine o'clock and crossed the Fournois. As the weather was cold and the water deep and swift, the teams, horses and young men forded, and the women, children and old men crossed in the boat. Etotahoma's cart was brought up and repaired. Encamped at about 4 o'clock. 9½ miles.

The land to which they were going—the so-called Indian Territory—would in due course come to be called Oklahoma, the land of the Red People. In the early nineteenth century, it was in effect the federal government's Indian dumping ground—immense and unpeopled expanse of billiard-table-flat prairie, lush green pasture in the east, dun-colored rough grazing in the west, the land basically surveyed and mapped, and now held in trust, in the eyes of the government, for the American people as a whole. It was the natural place, in Washington's view, for the reestablishment of the westward-trekking tribe-nations of the southeast. Therefore, in the late 1830s, certain massive sections within the territory became organized and settled as the Cherokee Nation or the Creek Nation or the Land of the Seminole. Other, smaller tribes were settled there from smaller—though equally tearful—trails of expulsion, and so there were lands devoted to the Kickapoo, the Potawatomie, the Ponca, the Iowa, the Sac, and the Fox.

But then in the late 1870s, once the territory had settled itself into some sort of pan-Indian Elysium, a Cherokee lawyer named Colonel Elias Cornelius Boudinot noticed something. There were two million acres of land sandwiched in an irregular T-shape—bounded on the east by the Kickapoo country, on the west by those square miles reserved for the Cheyenne and the Arapaho, and sandwiched between a part of the Cherokee country to the north, and the South Canadian River and the Chickasaw people to the south—that had been quite improvidently left "unassigned." No Indian tribe, nation or band had thus far been settled there, nor were there any plans for their doing so. The land was empty, still belonging to the people at large, in theory held for them in trust by a blissfully unaware government fifteen hundred miles away. It was a stunning discovery. Not surprisingly, once the news was out that two million acres of prime American real estate was there for the asking, every hopeful white man—in America and beyond—who was wanting land for himself, his family, and his future suddenly sat up and paid attention.

The idea of the land run of 1889 was born. A cry went up in the press across the country: "On to Okla-

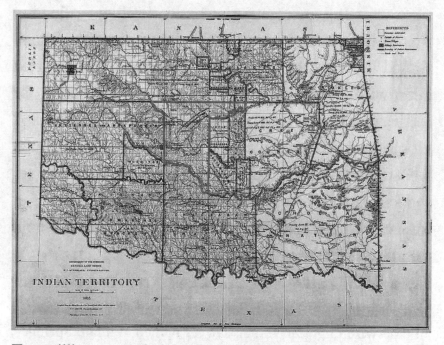

Two million acres of prairie land had been left unassigned to any Indian tribe by the planners back in Washington. They were free to be settled to anyone who might care to stake a claim—after winning a race to do so.

homa!" To the government in Washington, having no more Indians to round up and send away and stuff into the territory, it seemed an eminently reasonable idea—and moreover, the encouragement of migration of still more white people into the midst of an erstwhile Indian-dominated territory would, in the view of the capital, be no bad thing.

It was then agreed: a rider was inserted★ into the 1889 version of the Indian Appropriations Act—a piece of boilerplate legislation that dealt from time to time with Indian reservations and the like—allowing that at high noon on April 22 of that year, the Unassigned Lands, the T-shaped wedge of land buried deep within the Indian Territory, would be open for settlement. The basic terms of President Lincoln's 1862 Homestead Act would apply: for merely a registration fee anyone who staked a claim could have a quarter section, 160 acres of virgin land and, after working it for five years, could win title deed to it to have and hold for always.

★ Curiously, the congressman who inserted the rider to the act, an otherwise forgotten Illinois lawyer named William Springer, went from being a keen proponent of white settlement to becoming an ardent champion of Indian rights, fighting—but eventually losing—a case that tried to restrain government rights to break treaties with the Kiowa, the Comanche, and the Apache.

Like thunderheads on a humid Oklahoma afternoon, the crowds began to rise up and gather immediately. By late April they had gathered in their thousands, most of them when awake set to gazing hopefully across the cavalry-guarded boundary line and into the Unassigned Lands—gently undulating prairie grass country, with patches of April flowers: Johnny-jump-ups, little bluets, and spring beauties. The petals were so richly strewn on the ground that one writer suggested that the gods had left confetti.

One group of would-be settlers had left Kansas behind some days before the deadline and were now camped out with permission on Cherokee land, planning to ride south and, for want of good compasses, to follow the new railroad tracks and cross into the Unassigned Lands. The Atchison, Topeka and Santa Fe line to Texas had been completed just a year before, filling the growing need to transport cattle and hogs between the packing houses of Chicago and Gulf of Mexico ports, principally Galveston. Just south of the Cimarron River was the first train stop built inside the Unassigned Lands, a hastily thrown up way station called Deer Creek, which consisted of a single wooden platform alongside a water tower that could be hand-filled from the rivulet. In mid-April, anticipating the settlers' interest, the federal government had built two small structures close by: an army tent with the words *U.S. Post Office* burned onto a board suspended at the entrance, run by an Irish migrant named Dennis Flynn; and a rather larger, more solid-looking wooden structure that housed the U.S. Land Office. Meanwhile, the railroad bosses peremptorily changed the name of this station to honor their line's former chief attorney, John Guthrie, who was currently a judge in Topeka. So, by that designated Monday morning, a freshly painted sign proclaimed this otherwise deserted spot on the prairie to be the future site of Guthrie, Oklahoma. The crowds beyond the borderline were already familiar with the name: ask anyone where they might be found once the gun had sounded, and they'd reply with a roar: *We're going to Guthrie.*

All morning the post office and land office officials waited, drumming their fingers, twiddling their thumbs, in the warming April sun. It was quiet, with just a faint breeze as well as the occasional sighting of a grouse, a quail, or a black-tailed ground squirrel (known locally as a prairie dog), and every so often the chirp of an unseen bird. On all sides the prairie stretched out to the horizon, slightly undulating in parts, with shallow valleys marking limpid clearwater streams fringed with black oak and cottonwood trees.

There was no town, yet. But its borders were all

set out, however, scrupulously measured and marked according to the various land ordnances by which the country had been mapped over the previous hundred years. The township and range lines were all known and speedily drawn onto plats, large-scale wall maps that had been brought across by train from Washington and were now hung up on the Land Office walls. The all-too-familiar grid system—conceived as American Public Land Survey System, established in 1785, more than a century before—applied here in the Unassigned Lands, just as it had in Ohio, Alabama, Illinois, and virtually everywhere west of the Appalachian ridges.

All these rectilinear lines had their origins, their official point of beginning, at a granite monument on the banks of the Ohio River which still stands today, just outside the old broken-down pottery town of East Liverpool, Ohio. The obelisk identifies a nearby spot from which all of western America has been surveyed and measured ever since. Probably no one that day in Guthrie was even aware of the existence of East Liverpool; but those back in Washington who kept track of these things and who savored the true significance of East Liverpool knew well enough the transformative importance of the survey marks outside and inside the hitherto uninhabited settlement.

And Guthrie was indeed about to be transformed.

Anxious men checked the time. All of the Santa Fe workers would have their official railroad pocket watches, made by the Ball Watch Company of Cleveland. They looked at one another as noon struck. If guns had sounded up along the borderline, Guthrie was too far from everywhere for the officials there to have heard them. All was silent. An hour went by. Then, just before one thirty, two things happened.

A low growl, a grumble and a clang of iron against iron, and then with a snort and a whistle and a screech of brakes and billowing gouts of smoke and steam, a southbound train pulled to a halt at Guthrie station. With the slamming of doors and much yelling and war whoops, hundreds of passengers poured out of the carriages, pulled their luggage out and onto the platform, and lugged those bags and themselves down onto the thick meadow grass.

And at almost exactly the same moment, a thunder of hooves could be heard from the wilder grassland beyond: silhouetted against the northern sky like Zulu warriors breasting a gentle rise appeared rows upon rows of horses and swaying wagons, with hundreds and thousands more men and women, all who had ridden furiously down from the borderline. Most of the horsemen looked from a distance like medieval knights off to a joust: closer inspection revealed that they were carry-

Thousands who had massed on the northern border of the Unassigned Lands rode off furiously once the noon signal had been sounded on April 22, 1889. Within hours they had staked claims, and on payment of a five-dollar fee, the land was theirs.

ing long sticks of willow, with white flags and pennants fixed to their upper end. Claim flags, ready at hand to be put to good use down in the red dirt.

Thus was prairie silence suddenly broken, the sigh of the wind replaced by a great growing rumble of noise, a roaring cacophony of hooves and cries and whip cracks—and gunshots—and the yells of enraptured excitement. Thousands of people were now mill-ing around quite frantically, peering into the middle distance, eyes shaded against the sun, wheeling their horses and galloping off hither and yon, flagpoles at the ready, looking for suitable land to take and on which, in a trice and the blink of an eye, to settle

You jumped from the horse, plunged your stake into the ground, maybe wrote your name on a board and built up a pile of bones and buffalo chips to leave the board on, confirming the claim and your name to it. You noted the survey number on the nearest marker post, if it could be glimpsed through grass that was often taller than your horse's head. Next, with the location of your chosen lot, its coordinates, committed to memory—and it had to be a good memory, since a lot could be designated according to the survey rules just as, for example "the SE quarter of Section 8, T16N, R2W"—off you cantered to the Land Office. Here, after waiting in line with scores of others, you eventually completed a government form, Number 4—963, a "Homestead Affidavit, Sworn in Guthrie, O. T.," in which you solemnly swore that you did not own 160 acres of land held in fee simple in any other state of the union, that you were either the head of the family or over twenty-one, that you were a native-born citizen or in the process of becoming a naturalized ditto, and that you had paid the necessary registration fee of five dollars.

After which, in a heartbeat and with a stroke of a pen, the 160 acres of land at the South East corner of Section 8, Guthrie, O.T., hitherto the sovereign property of the republic, its recent purloining from the native inhabitants long forgotten in the heat of the moment, became indubitably and permanently or for as long as you wished, yours. You were now—providing no one sued you for jumping the claim or having an overlapping interest, which happened all too often, and understandably given the chaos of the day—the owner of a piece of the United States of America. It was but a tiny, near invisibly small fraction of the whole 37 billon acres that made up the continental republic, but nonetheless it was yours, to have and to hold, and on which to build a part of this brand-new city of Guthrie.

And build it, did they ever! By late afternoon the prairie was an ocean of tents, the line at the Land Office half a mile long. The newcomers unloaded from their wagons their spouses, children, and servants; or if they had more heroically arrived by horse or by train, they took from their saddlebags or their attaché cases more frugally selected possessions: plates, jugs, spoons, boots, nails, pieces of wire. More trains arrived, disgorging still more people. By nightfall, and with a few glimmering candle lanterns in sputtering lines north and south, east and west, so the prairie suddenly had the appearance of a city in the making. There was little water, there was no coal, no toilet facilities. A few enterprising folks had erected larger hotel tents and trestle tables; they were selling food at exorbitant prices and offering shots of bad whisky to those who had not brought their own. Most settlers had guns, but so far as is known, there were no serious outbreaks of violence. There was a common recognition that the emptiness of the prairie—and not those others who were there to help build a city—was the common enemy that had to be overcome. As the dark enveloped the settlement and the stars shone down from the vast midwestern sky, so the City of Guthrie had been brought formally into being, its new inhabitants promptly falling into a bone-weary sleep.

By morning, once the cooking fires were lit and with the ham-and-eggs and coffee smells of breakfast wafting over the site, someone rode a horse northward, between one line of tents, and suggesting as he did so the making of the first road in town: Division Street, he shouted out its name. This would delineate the halfway point, with all streets to its west to be named thus, all to

the east named otherwise. And the first of these bou-
levards to be named was Oklahoma Street—born with
its north–south cousin in late April 1889 and crossing
at right angles then as it still does today, a century and
more later.

A writer named Hamilton Wicks was up early. He
wrote an essay for the September issue of *Cosmopolitan*
magazine, then a rather different journal from today.

I strolled up on the eminence near the land office,
and surveyed the wonder cyclorama spread out
before me on all sides. Ten thousand people had
"squatted" upon a square mile of virgin prairie that
first afternoon, and as the myriad of white tents
suddenly appeared on the face of the country, it
was though a vast flock of huge white-winged birds
had just settled down upon the hillsides and in the
valley.

The speed of the city's transformation beggars belief. At dawn on that first Monday the settled population of Guthrie was precisely zero. By ten that evening
the population was not the ten thousand that Mr. Wicks
had suggested, but by some estimates more like fifteen
thousand. Even allowing for families, it can fairly be

said that one settler every three seconds would lay claims to land in Guthrie, Oklahoma, that April day. No other city in America would ever grow so fast.

Twenty days later, the ramshackle tented community that the settlers had established had transformed itself into a proper town, with a telegraph connection to the outside world, with fully assembled buildings (some still standing today), a chamber of commerce, and no fewer than three competing daily newspapers. Inside two months, Guthrie had piped running water; eight weeks after that, some of its houses, offices, and unpaved streets were lit with electricity. By the end of the year Guthrie was a full-fledged community, briefly destined to be the capital of the state, once statehood was declared in 1907.

This last was a dream soon extinguished by the ambitions of Oklahoma City thirty miles to the south. But Guthrie had its brief moment in the sun—there would be a struggle over the forcible removal of the Great Seal of Oklahoma from the bank safe in Guthrie to the new government headquarters in Oklahoma City. Lawsuits, court orders, and the threat of police force would all be involved in wresting that huge device from the hands of the proud burghers of Guthrie.

Since its decapitation the city has been somewhat sidelined—it has a bluegrass festival and a steer roping rodeo, and one of the largest Scottish Rite Masonic Temples in the country. But its real gems are its first buildings—a cluster of some two thousand highly anachronistic structures* that rise where otherwise grain elevators and undistinguished suburban tract housing should normally appear. The city is a mausoleum of the American land saga, a memorial to Victorian pioneer history, a somewhat melancholy reminder of the lengths to which many brand-new Americans would go, in order to acquire that most precious of com-

---

* Many architects of the time would come and try their hand in the building of Guthrie—most notably, a Belgian named Joseph Foucart, who left Europe (where he had designed the Paris City Hall and the Brussels Grand Hotel), and came to America with a view to building "Castles on the Plains." Seventeen of these monsters of sandstone and granite (eight of them still standing) he built in Guthrie. Noble structures—banks, city halls, publishing companies—were his *metier*, and one of the reasons so many remain in Guthrie and cities around the prairies is that he built them so well they were nearly impossible to demolish: they simply "defied the wrecking ball," as one newspaper wrote. Many find them irredeemably ugly. Developers also came to Guthrie, which was not a city entirely given over to would-be yeoman-farmers. The best-known developer and town champion was a Canadian, Hobart Whitley, who went on westward from Oklahoma and bought large tracts around Los Angeles—including one that, on the suggestion of his wife, he called Hollywood.

The town of Guthrie lays claim to being the fastest-growing city in America—no one lived there on the morning of the twenty-second of April, but by sundown, it had a population of fifteen thousand. And soon, a cluster of impressive Victorian buildings.

modities, that which, to paraphrase Margaret Mitchell, is seen as worth fighting, dying, working, and living for, because of its enduring qualities, because it lasts.

The Native Americans whose lands these once were have not forgotten, nor probably ever will. But such anger as they might justly feel has long ago ebbed, and it just simmers in the far background. The notion that the United States could ever have had the effrontery to call these acres "unassigned," when in truth the republic never possessed the moral authority either to assign or unassign them, here or anywhere else, is mostly forgotten today. The world has moved on, even if in the small microclimate that is Guthrie, Oklahoma, the world appears on the surface to be very little different from the days just after the Oklahoma Land Run was completed.

4

The Land and the Gentry

And then there are those whose acquisition of land they believe to have come about through ancient right.

These are by and large Europeans and, most egregiously, they comprise a large number of Britons—English, Scots, Irish, and Welsh—who occupy a special place in society, wielding legislative, economic and social power still, to a degree that many quite reasonable onlookers currently think of as anachronistic and improper.

These few hundred men and women are the landed gentry, and, together with the dukes and marquesses and earls and viscounts and barons and assorted lesser titled British subjects, they currently own impressively large chunks of the British realm. Fully a third of the

English countryside, for instance—13 million acres of the 37 million that make up the totality of the surface land area of England and Wales—belong to private firms or land-rich families of often great antiquity. Of Scotland's twenty million acres, almost half are controlled by some eleven hundred wealthy landowners. Between them the country's twenty-four non-royal dukes own more than a million acres. Much the same degree of individual abundance obtains in Northern Ireland's six counties. Whether such a disproportionate amassment of land in the hands of so very few is right or justified or prudent, wise or sensible, is a matter for a later chapter.

As to how such families might have acquired that land, the stories are generally befuddled by antiquity and legend and have acquired a patina of long forgotten mythology. The finer gradations of the British class system, which despite all efforts at meritocratic evolution has still not yet wholly vanished, and certainly not in the farther reaches of the countryside, recognize that the gentry—whose male members are literal *gentlemen*, a term of much subtlety little known beyond England—stand both at the very summit of society, and act as the great pillars that support the national establishment. These are the "old" families of England, the landowning classes, gentlefolk untouched by titular reward, generally unwilling to be involved in such vulgar matters as politics or—heaven forbid—*trade*, which might taint their social sanctity. In times past— when landownership gave one the exclusive right to vote—these were the ruling elite of Britain, those who by great good fortune or by the courtesy of kings had managed to amass what in Britain is still the most stable and most respected form of wealth, acreage. They own land; their offspring—in most families, the oldest male inheriting the land and sufficient of a fortune to run it, of course—becoming in descending order of birth members of Parliament, soldiers, diplomats, barristers, and, in the case of the youngest or most dim, parsons in the Church of England. Or, longer ago, functionaries sent out to run the more distant reaches of the empire.

Matters are changing, though. Change and decay have taken their toll. Few enough these days enjoy the life of, as an example, the country gentleman encountered some while ago near the border village of Caledon, County Tyrone, in a currently defiantly British corner of Northern Ireland. He lived in a castle, modest by the granite extravagance of most Scottish standards, but comfortably situated in some 29,000 acres of lush meadowland on the banks of the River Blackwater. An eighteenth-century ancestor returning from an evidently moneymaking venture in Bengal purchased the estate for £9000 in 1772, anticipating an annual income from rents of some £7000—an ample enough sum to guarantee his becoming one of the country's leisured few. The family's effortlessly untroubled lifestyle evidently continued for at least the subsequent two centuries, for when over breakfast one day it was asked what he did on an average day, this Caledon gentleman chortled with amusement. "Do?" he thundered. "Do? What on earth do you mean?"

He then explained that he did as little as possible because, as he spluttered, "Nothing in my life is ever urgent." On being pressed he allowed, after thinking for several minutes, that on that particular day he might perhaps venture into the village to order a case of sherry to send to his mother across in London; and then recalled that on one of his many farms there was said to be a new child, whom he might visit, if the weather was clement. Otherwise, "the nice thing about owning a great deal of land, and having tenants who farm it, is that everything more or less looks after itself. There really is very little that I have actually to do." And with that he snapped open the latest copy of *Country Life* and began reading an article about some rare breed of pig.

However serenely enjoyable such a life may be, the gentry is today not exclusively of the rentier class.

Anyone who sits on enough land is likely, thanks to the providence of geology, to have something of value underneath. And though in the past the landed gentlefolk have been generally disdainful of allowing the extraction of minerals from beneath their acres, some today do, and make spectacular supplementary fortunes. Matt Ridley, a prominent newspaper columnist who inherited a title and an estate of 15,000 acres in the northeast of England, happily allows the opencast mining of his parkland, providing that the dragline operators—whose clanking work can be vaguely heard through the thick walls of Blagdon Hall—clean up after themselves and pay him both a rent and a share in the revenue from the sale of the coal. A neighbor member of the gentry living in a vast country seat just to the north of the Scottish border is also fortunate to have had underground mines on his estate—and memorializes the profits they once brought by having a piece of antique silver mounted on an enormous plinth carved from a single two-ton block of coal.

John Maynard Keynes remarked critically of such people, guaranteed as they were an eternal income simply because land itself was finite and scarce and the populations requiring it becoming ever more numerous. He cared little for the gentry's argument that they were custodians of the land and creators of the landscape that was, in Britain at least, so uniquely and transcendentally beautiful—a landowner planting oaks that he would never live to see mature, but content that by doing so he would be helping to continue the charming aspect of the view. Keynes retorted only that he thought euthanasia the best solution for the gentry, the country all the better for getting rid of them.

Though time and taxation may be doing as he wished, slowly winnowing away the gentry's numbers and diluting their influence, the matter of how the land was apportioned in the first place intrigues. The Caledon family's Irish land was acquired mainly by simple purchase, the money made in eighteenth-century Bengal having been used to buy acreage that the local bishop, in need of cash to shore up his creaking old cathedral, thought surplus to his pastoral needs. But purchase was not invariably the way in which the ancient landed families acquired their territories. Other tracts came into the hands of members of the gentry by more convoluted means, and in many cases so long ago that no one could be entirely certain by what right their ownership originally commenced.

The most comprehensive compendium of long-ago ownership in England is, most famously, Domesday Book. This is in fact an elegant pair of volumes: Little Domesday, a summary of 475 vellum leaves, and Great Domesday, the more finished product with a more tightly written 413 vellum pages. Both volumes are kept today under formidable security in Britain's National Archives. They comprise in essence the published result of the first-ever inventory, painstakingly written in one medieval scribe's Latin hand, of England's landholdings and landownership almost a thousand years ago.

The survey was commissioned by King William I, William the Conqueror, who in 1066 at the Battle of Hastings had famously vanquished the very last of the Anglo-Saxon monarchs, King Harold. His survey was conducted in 1085, and evidently quite swiftly, for the resulting book was first published only in 1086, with the presumed purpose of allowing the new king to calculate the taxes that would be due on the lands that were privately owned around the nation.

The great Victorian parson-historian John Giles* made a popular translation of the *Anglo-Saxon Chron-*

---

* This exceptionally polymathic figure had a son, Herbert Giles, who went on to invent with his colleague Thomas Wade the once much employed Wade-Giles system for the transliteration of the Chinese language—Mao Tse-tung rather than today's Mao Zedong.

The two Domesday Books, presenting a comprehensive handwritten inventory of all the land in eleventh-century England, are among the country's most treasured possessions, and still have legal standing.

icle, which described in rather greater detail how the Domesday survey was performed, and of King William's role in seeking to ascertain

how the country was occupied, and by what sort of men. Then sent he his men over all England into each shire; commissioning them to find out "How many hundreds of hides [an ancient measure equivalent to some 120 acres, supposedly enough to sustain a household] were in the shire, what land the king himself had, and what stock upon the land; or, what dues he ought to have by the year from the shire." Also he commissioned them to record in writing, "How much land his archbishops had, and his diocesan bishops, and his abbots, and his earls;" and though I may be prolix and tedious, "What, or how much, each man had, who was an occupier of land in England, either in land or in stock, and how much money it was worth." So very narrowly, indeed, did he commission them to trace it out, that there was not one single hide, nor a yard of land, nay, moreover (it is shameful to tell, though he thought it no shame to do it), not even an ox, nor a cow, nor a swine was there left, that was not set down in his writ. And all the recorded particulars were afterwards brought to him.

The convulsive events of the Norman Conquest—Domesday being among its many legacies—profoundly changed the makeup of England: the rule of the Anglo-Saxons was ended forever, and the ground was steadily prepared—socially, legally, and linguistically—for the evolution and eventual creation of today's society. And particularly so far as land was concerned. The Norman kings—there were three of them, eight if one counts all the confused gatherings of monarchs who ruled from London until the much more truly *English* Plantagenets firmly established themselves on the throne in 1216—were both extraordinarily rapacious in grabbing land and gaudily generous in offering it to their cronies.

These fortunate figures, distant ancestors of many of the thousands who make up today's gentry, were a tightly knit coterie of some two hundred of the most favored barons, courtiers, and members of the Catholic clergy, together with such knights as proved themselves willing to ride off to the European mainland to wage wars on the monarchs' behalves. The beneficiaries of the kings' largesse had good reason to be content. Less happy, however, were those from whom the lands had been taken. And yet these unhappy owners had themselves succeeded previous owners who were in turn victims of plunder and greed, and those too had taken land from still earlier owners, with a chain of land-related purloining proceeding ever backward into the dark depths of ancient English time.

The Normans were tough and cunning fighters, and they took more or less as they pleased. For decades after the invasion, the Breton broadsword and the Norman longbow were employed to French advantage all across the acres of newly occupied England. One by one manors and domains and wapentakes and hides that hitherto had been owned by the Danes and Norwegians and Saxons fell into their hands.

This antecedent Anglo-Saxon England had for its six hundred years of existence been a confused arrangement of petty kingdoms—Wessex, Mercia, Northumbria, and Kent among them—that were created after the departure of the Roman colonists in the fifth century. It was during this time that the country would become more organized and eventually united, under the name of England, to be ruled by a succession of English monarchs.

During its prime, Anglo-Saxon England did, however, create a system of landownership, and it is fair to say that during the period leading up to the Norman invasion just about all the land in Old England was owned by someone. The word *acre*, so central to the concept of land and landownership, was born during these times, and it remains—along with *bread* and *earl* and *half*—one of the oldest recorded words in the language, indicating how important concepts relating to food, society, and arithmetic—and land—were to the earliest English.

Formal concepts relating to land were adopted, with the various kings exercising the rights of ownership and their assistants, the thanes (or thegns, as they were first spelled) being given rights of tenure and in most cases actual written charters to allow these rights to be passed on to their descendants. It is believed that around four thousand thanes were tenants of kingly land—known as *bookland* if the thanes held a charter, *folkland* if not—in tenth-century England. Most of these or their successors would have then lost their holdings and their security once the Norman barons had vanquished their monarchs and stolen their tenanted parcels away.

Go backward still further, and matters become somewhat more vague. Before the confusing times of the Saxons, and during the reigns of the Danes and the various interregnums and rules of other Scandinavian invaders whose encroachments on the British Isles made for so many decades of battling and mayhem, there were the Roman colonizers. Their four centuries' rule of *Britannia* was at least disciplined and organized, if not exactly allowing for self-determination by the inhabitants. Scholars believe that the entirety of the conquered country—right up to Hadrian's Wall, which had been built in the far north of England to keep the bellicose Picts at bay and to help preserve order even at the most distant reaches of the empire—was regarded as imperial property, but which could be and indeed was occasionally given out to some local tribe as either reward or inducement.

But the clues are largely inferential. Land, it is believed, played a signal role in one of the great episodes of early English heroics. Boadicea—Boudicca, Victoria; her names are many—the great English warrior-queen of the Iceni who rebelled against the Roman rulers in AD 60 (and burned down most of Londinium in the process), is thought, for example, to have staged her insurrection because the imperial governors had confiscated land hitherto given to the Iceni for some unspecified reason. That the Romans had given the land★ in the first place suggests to scholars today that Rome firmly believed itself to be the owner of all En-

---

★   Rome's chief financial agent in Britain at the time, Decianus Catus, claimed that the gifts were actually loans—a cunning trick that triggered the Iceni rebellion in the first place. It became a vicious affair: for her impertinence in challenging imperial rule, Boadicea was whipped and her two daughters raped by centurions, after which she raised an Icenian army, famously sacked the garrison towns of Londinium, Camulodunum (Colchester), and Verulamium (St. Albans), and perished only after an immense battle in English Midlands in which she employed her chariot with, supposedly, curved fixed scimitar-blades whirling from its wheels. Tens of thousands died; she was poisoned; Rome almost abandoned its faraway British colony—and all because of dispute over a few thousand acres of eastern English land.

glish land, and could distribute it, or rent it out, as it saw fit. Such assumption spawned the greater idea that Roman rule over England existed, and was absolute.

Prior to the Romans' four centuries of generally stabilizing presence, the idea of the ownership of early England becomes supremely vague, and assumptions about its arrangement tentative in the extreme. We know of the existence of the British Isles' various tribes—tattooed and woad-painted Celts—but little of their attitude to land. Except it is known they ploughed, with two-oxen teams hauling the primitive ploughs and tracing patterns of furrows over the friable and stony earth of the untilled countryside. Where a plough team completed a line and turned back, left behind was an eruption of piled earth—and after a time this pile of earth became a crude fence, the demarcation, a line in the ground that divided one pattern of ploughing from another, one area of personal settlement from another, one owned field from the owned field of a neighbor.

And such, it seems to those archaeologists and anthropologists who have contemplated the finer relict patterns of the English landscape, is how ownership began. It started with the lines of furrows left behind by the ploughs of neighbors, probably fourteen hundred years before the birth of Christ, thirty-five hundred years ago.

In those ensuing years, the concept of landholding in England, which had so considerable an influence on the patterns and practices of landholdings all around the western world, took shape, developed, and evolved. Ploughed field furrows became eventually substantial stone walls and Roman villa building; Saxon kings divided their fields among their faithful thanes; Norman monarchs devised the idea of kingly ownership and the feudal granting or gifting of real estate; and from the eleventh century, laws and policies spread ripples of consolidation and control that have left millions of acres of today's England in the exclusive ownership of a *corps d'elite* of (mostly) men, who with their families belong still today to a titled and untitled aristocracy that owns—or to be exact, holds estate in—so many millions of acres of modern England.

"This dear, dear land," complained old John of Gaunt in Shakespeare's *King Richard II,* "is now leased out . . . Like to a tenement or pelting farm." Ownership of land brings with it a grave responsibility, not least when one considers the state of today's planet. Whether the planet's surface has been tended well, or ill, has over the centuries been to many a matter of profound and enduring concern. It was when people began building fences around it, enclosing public land for private purpose, and when others, more cruelly, began dismissing landed tenants and replacing them with livestock—both acts being seen not as social vandalism, but as means of improving the quality and the usefulness of the land—that matters first started to go awry.

PART III
Stewardship

1

The Tragedies
of Improvement

The law locks up the man or woman
Who steals the goose from off the common;
But lets the greater villain loose
Who steals the common from the goose.

—ANON.,
in *The Tickler Magazine* (1821)

At the start it was all down to the ploughing methods and the field patterns of two Bronze Age
farmers. Many years later it came down to two ideas,
two movements, two series of long-drawn-out events
that were put to work on land that was by now fully
known, divided, managed, inhabited, and used.

Each in its own way was a movement intended to make the use of land more efficient and, for its owners, more profitable. The older of the two—enclosure—was a phenomenon generally backed by the state. The latter and more notorious—the more obviously cruel and generally Scottish experience known as clearance—was privately inspired. Both had profound human consequence. Both led to waves of migration: the hundreds of written enclosure acts persuaded millions to move from the countryside to live in towns, while the clearances compelled thousands to flee the wilds of Scotland and settle anew in the unrestricted emptiness of North America. More glibly, one can say that enclosures brought about the invigoration of great cities; the clearances helped to create Canada.

The enclosure of land, the enforced removal of a portion of the surface of the Earth from the common ownership of many to that of one or more private individuals, represented a revolution in the social order, a cataclysmic change like few others before or since. It had been going on without formal sanction for years—the earliest records show land being enclosed, often against the villagers' will, in the thirteenth century. In 1604, Radipole, a village in Dorset, was the first to have the backing of a Parliamentary Act—the first of almost five thousand specific acts for towns, villages, moorlands, and pastures all around the country that would be passed between then and the early part of the twentieth century, each under the terms of enabling acts which remain on the books and are still theoretically in force today.

Enclosure of an even less formal nature is still more ancient—with many believing it to have been occurring in England as far back as the late Bronze Age four thousand years ago, at the time of the previously mentioned Deverel-Rimbury people, whose field patterning set in train the idea of land demarcation and its private ownership.

The earliest lands back then were covered with primeval wildwoods, and archaeological evidence suggests that even as far back as the Stone Age these woods might have been partly felled and ravaged to provide rough grazing for captured animals—leaving behind a landscape of gorse and furze and bracken and relict stumps and fresh undergrowth. Come the Bronze Age, and its sudden semisophistication in farming methods, and the previously roughly grazed cattle were corralled into what one might term *ranches*, with islands of cleared meadow happily endowed with good grass and flowing water. These grassy areas were specifically reserved for arable farming—parcels of land decidedly not to be trampled upon with hooves and

horns, and so enclosed from the less agriculturally congenial stump-laden pastureland.

No concept of ownership had yet evolved—but the
idea that some especially motivated individual might
assume direct responsibility for, and an interest in, a
specified piece of enclosed land was born at that moment. The conventional notion holds the first true
enclosures were in Tudor times—Radipole in Dorset
being the classic example, though technically its taking
in 1604 was of course a year into the reign of the Stuarts.
But this is only partly true. The process, where land is
concerned, of having and holding may have been informally organized, but it had been occurring nonetheless
for centuries, suggesting that the habit of taking over
land for oneself has been an inherent human trait for a
very long while.

Informality had reigned for many previous centuries.
Customary law, which long predated the thirteenth-
century invention of precedent-based common law, and
which varied from village to village in Old England,
regulated—if regulated be the word, for in truth it was
much less formal than that—the assignment of village
land throughout the Middle Ages. And assignment was
often sorely needed—for while a villager might have
an acre or two of land that he could cultivate, he and
his family could barely survive without also having a

cow or two or a pig or three or a gathering of sheep or chickens—and yet he could not allow these animals to graze on the land that he was cultivating, in case they ate all the crops and vegetables that he was trying to grow at the same time. He would need somewhere else to graze his livestock—and also somewhere to fell and split his firewood and collect his turf—and for these purposes custom allowed him and his neighbors the right to do so on the land nearby that was used *in common*.

This was the so-called wasteland, lying on the outer edge of the village, occupied by no one and thick with cattle, trees, and rough and untended grass. This was for the peasantry an accepted aspect of English village life, long established and settled according to customary law. The common man had rights to this common land, this commonly held land. And they were rights that, though varying in the detail from place to place, were both recognized and, if all around were reasonable men, legally inextinguishable.

Except that they could be extinguished or traded *by agreement* with neighbors. The villagers might all agree, for example, that the common land on which they all exercised these rights might somehow be made better use of in some way—by this part being reserved for cows, that for sheep, this corner set aside for all the village pigs, or this copse left for the growing trees and that area being left as rough grassland. If, in other words, there was some sense of order to be brought about on the hitherto free-for-all common. Part of the commons could be *enclosed*, to use the term of art, and by mutual and commonly accepted consent.

This is how it was accomplished, and for many hundreds of years reasonably happily so. But as history illustrates on all too many occasions, with education and sophistication comes awareness—and by the fifteenth and sixteenth centuries, a number of educated and aware men and women started to have arguments over the perceived inequities of the common land system. People started to cry foul over the readjustment and reassignment of various tracts, and radical leaders empowered by fine oratory and impelled by what seemed a *prima facie* injustice—men like Jack Cade in Kent, Robert Kett in Norfolk, and John "Captain Pouch" Reynolds in Northamptonshire, and groups like the Diggers in Surrey—found themselves able to assemble angry crowds who gathered to protest that this field had been unfairly distributed or that fence spoiled the common access to a particularly fertile meadow. Mobs set about destroying fences and filling in ditches and driving their sheep into forbidden pastures (or, as with the Diggers, illegally planting carrots on a hill south of London).

Through the early part of the eighteenth century, this unease showing itself was born from more than just the simple freelance acts of enclosure that so troubled the peasantry of the day: rather there was a distinct feeling of an inchoate unease, a gathering impression that the country had become awash in new-fangled ideas, part of what we now know to have been the beginning of the Enlightenment. There were new and unsettling developments in farming techniques, the introduction of machinery and of four-crop rotation methods, which we in retrospect now recognize as the Agricultural Revolution. There were hints of the coming of the Industrial Revolution too, a revolution that would soon sweep like a gale through all of English society and would massively enhance the rise and role of cities, which would lure workers in vast numbers away from the countryside.

The unease of the times parallels to a degree the same kind of bewilderment at the rate of change in society that so clearly afflicts twenty-first-century humankind. A vague but gathering sense of anxiety was spread more by taproom gossip and from riders of the stage lines, slowly and fitfully. However, it gave the rural poor—especially once they also saw fences rising all about them—the idea that matters were taking

place that were somehow not consonant with the supposed idyll of arcadian pastoral life. And so, in many
scores of villages there were riots and violence and a
scattering of deaths. The government responded with
curfews, imprisonment, charges of treason, and not a
few hangings, and a growing alarm was felt in London
that the situation in the countryside might be getting
out of hand.

After much debate and argument, the government
in its wisdom set about laying down rules. With the
intention of making certain for everyone's good that in
the matter of enclosure, ownership become a known
and formalized quantity, and there was paperwork and
the drawing of maps and the making of deeds and the
issuance of titles that would formalize and make somewhat more defined the whole fast-changing edifice of
the private ownership of England's solid surface.

And thus came the first formal and officially approved enclosure of common lands, in and around the
Dorset village of Radipole. For the next three hundred
years it was ruled that the formal approval of Parliament was necessary for this very particular kind of
land seizing. The process of winning the acquiescence
of the government in London was both costly and
time-consuming for whoever applied, but it followed
more or less the same pattern in each case.

It would begin with a formal request from the proprietor, as the ultimate owner of the land was known, that his land be enclosed and the rights of the commoners who customarily used it for livestock or wood or turf collecting be extinguished. A private bill asking for this would be drawn up, with the details laid out for all members to consider. Parliament would then examine the request on its face: Was the enclosure justified? Would farming become, as usually claimed, more efficient? Would the land become more productive? (With the English population now growing fast, more and more food was needed from the nation's farmers.) And then again—who would be affected, who adversely, and how badly—and who might benefit, and would they do so in a reasonable or an unduly generous manner? And what rival claims might there be?

Parliament would then order that a parchment document giving notice of the possibility of local enclosure be fixed to the church doors in the parish, for the next three weeks, just like the banns of marriage. Once the bill passed such muster—at a time when perhaps not all churchgoers had full command of written English—three commissioners, men of good standing and with a presumed neutrality in the matter, would be dispatched to examine the proprietor's request and such commoners' claims as had been thrown up, and then try

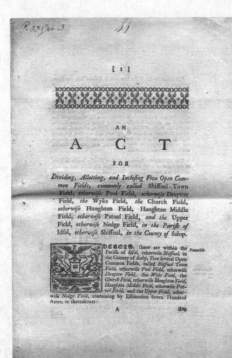

A formal printed Parliamentary Enclosure Act would be tacked to the church door to be read by all. After a decent interval, it permitted the local common land to be fenced off and turned into private property.

to determine if indeed the enclosure served the public good or not. And once the commissioners' report was in, so the bill would be debated, voted upon, and, if successful, translated into a formal act of Parliament—meaning that the enclosure itself, of this particular village or parish or common field, would be the settled law of the land. Challengeable still in the courts, but effectively a done deal.

And as mentioned, it was a deal done nearly five thousand times through the eighteenth and nineteenth centuries, as bill after bill was presented in Parliament and millions of acres of once commonly used land came to be ringed by wattle and wire, and fields, meadows, and woodlands were swept into private hands. The controversies attendant on the very idea of enclosure have not subsided over the years since, and the complexities of both its initial justification and its often dire consequences continue to exercise the minds of academics and social commentators to this day. Its complexities are indeed very real, and the social historian E. P. Thompson issued a warning, less heeded than it should be, that "a novice in agricultural history caught loitering in those areas with intent would quickly be despatched."

So far as justification is concerned, the official reasons can be inferred from the long title of the enabling act that Parliament passed in 1773, that "inclosure" was a process designed "for the better Cultivation, Improvement, and Regulation of the Common Arable Fields, Waste and Commons of Pasture in this Kingdom." The unstated reason, however, was suggested by an infamous paper published in *Science* magazine very much later, in 1968, which had a title that has since become a catchphrase: "The Tragedy of the Commons."

The essay's author, an American ecologist named Garrett Hardin who was a devout anti-immigrationist and eugenicist and a believer in the dangers of overpopulation, declared that commonly used land would inevitably be badly used land, because people were greedy or careless, wouldn't cooperate or take care of their land, would push as many of their own cattle to graze on fields that were already overgrazed, would take more than their fair share, and so on—would, in other words, ruin the bounty that God and Nature had so generously offered to them. Hardin quoted William Forster Lloyd, a Victorian divine and Oxford mathematician who in 1833 had asked:

> why are the cattle on a common so puny and stunted? Why is the common itself so careworn and cropped so differently from the adjoining enclosures? If a person puts more cattle into his own field, the amount of the subsistence which they consume is all deducted from that which was at the command of his original stock; and if, before, there was no more than a sufficiency of pasture, he reaps no benefit from the additional cattle, what is gained one way, being lost in another. But if he puts more cattle on a common, the food which they consume forms a deduction which is shared between all the cattle, as well that of others as his own, and only a small part of it is taken from his own cattle.

This, as Hardin described it, was the true tragedy of the commons, and the only sure way to avert it was to make the land private. "Common ownership remorselessly generates tragedy," he wrote, and then with a final flourish, offered a paragraph memorized by all who have strong views about enclosure and common ownership:

> An alternative to the commons need not be perfectly just to be preferable. With real estate and other material goods, the alternative we have chosen is the institution of private property coupled with legal inheritance. Is this system perfectly just? . . . We must admit that our legal system of private property plus inheritance is unjust—but we put up with it because we are not convinced, at the moment, that anyone has invented a better system. The alternative of the commons is too horrifying to contemplate. Injustice is preferable to total ruin.

Battle has been joined ever since.

The political left deplores the enclosing of land, believing it to have led to dispossession of the rural poor, the creation of a perpetually impoverished underclass, the cementing of a system of unfair distribution of wealth, the consolidation of the imbalance of landownership, which leads inevitably to the introduction of factory farming, the withering away of the pastoral idyll, and the ruination of all societies that have toyed with this most sacrilegious of notions.

The political right, on the other hand, believes enclosure to have led to immensely more efficient farming, to ever greater food production, and to a healthier, better-fed, and richer population, and to have set those countries that have been bold enough to do such a thing on the road to the sunlit uplands of democratic capitalism. Karl Marx denounced enclosure in *Capital*, claiming that by transforming feudalism into capitalism it made wage laborers out of peasant proprietors, and then ruined them by forcing them off the land as a result of enclosure-effected economies of scale. The subject is one that invites the electromagnetism of the doctrinaire, with little quarter then or today given by either side.

Whatever the principles at stake, the social effects of enclosure were widespread and undeniable. Those who lost out in the lottery of land redistribution drifted away from the countryside and, once the factories of the Industrial Revolution started belching out smoke, took work in the business of making and manufacturing. Cities suddenly began to expand, explosively; the villages emptied as thousands of one-way tickets were purchased by a population that was now never likely to come back. And similarly, when ships in Liverpool, Glasgow, Belfast, and Cork started firing up their boilers and lighting out for the heady promises of Boston, New York, Baltimore, and Montreal, so the unemployed and underused peasantry of England purchased passage tickets, more expensively, to begin new lives ordained by the fencing off of their old homesteads left firmly back at home.

There is a town in eastern England named Barton-upon-Humber. It suffered—or enjoyed—enclosure beginning in 1793, before which year its 5800 acres were divided essentially into six enormous fields, open to all the villagers to graze and till as customarily allowed. By the time the enclosing was all done, the town's land was fenced and walled and ditched and thereby divided into myriad smaller properties, and the names of the owners are inked proudly onto the map: James Richard now has 81 acres of his own land, George Uppleby "and Sarah his wife" have 156 acres, William Graburn 252 acres, Thomas Scrivener 71, William Holgate 89, and so on, scores upon scores of new owners or tenants, where before there was just a single owner and a good deal of ancient rights.

Perhaps not all of these new possessions proved affordable. Perhaps some of the new farmers were not so proficient, and their crops failed, or their cattle succumbed to murrain and blowfly. Maybe these men failed, and left. But some, possibly even most, remained, and now they had land, they had real estate, they had the fee simple ownership of acreages of the solid surface of England. Land which, as Winston Churchill would famously remark in 1906, "is a necessity of all human existence, which is the original source of all wealth, which is strictly limited in extent, which is fixed in geographical position—land, I say, differs from all other forms of property in these primary and fundamental conditions."

Seen through the eyes of a wily politician like Churchill, the enclosure legislation that was to wind down at the beginning of the twentieth century, its work effectively done, was to result in very much more than "the better Cultivation, Improvement, and Regulation of the Common Arable Fields, Waste and Commons of Pasture in this Kingdom." It was to help create and stimulate a slow-moving social revolution; and though it is seen by many as helping to increase the power of the very rich at the expense of the landless and very poor, it can be argued that it may have helped to foster an environment in which the middle class could own land for the first time—as the map of

Barton-upon-Humber and thousands of other similarly enclosed places will demonstrate.

Ordinary people, men and women with access to some modest sum in seed money—obtainable by loan, maybe, and from a bank that would in theory be only too happy to advance a small sum that would be secured by the inherent value of the land which was to be purchased with it—could now each acquire a modest parcel, and thereby start the process of the steady accumulation of capital that lies at the heart of a capitalist economy. Though some will suggest this to be wishful thinking, unabashed capitalists would argue that by way of enclosure, riches were on the horizon.

But up in Scotland, and at around the same time, a very different kind of land reallocation was taking place, and with very much more wretched and miserable results. While in England only a few nowadays recall and have settled views of the phenomenon of the enclosures, across the border in Scotland just about everyone recalls the events of which little good can ever be said: the clearances.

The most notorious practitioner of this peculiarly cruel kind of land stewardship had a statue raised to his memory and remains two centuries later a much loathed symbol of all that is blighted about the relationship between Scotland and England.

They call it "the Mannie," it stands in the north-east of the country on a ridge known in Gaelic as Ben Bhraggie, *the speckled hill*, and it can be seen from tens of miles away in all directions. From a long way off it looks like an enormous radio mast piercing the far blue hills of the horizon; come closer and you see it is in fact a great stone statue of a standing man on top of an immense stone plinth. Closer still and the man, his face of noble appearance, is seen to be facing to the east, his body is draped in classical robes. There is an oblong object, a notebook of sorts, in his left hand, carved leaves of bay laurel on his head, like a victor.

There are many things about this hundred-foot-tall red sandstone edifice with which angry local Scots— and, indeed, angered Scots around the world—still take issue. Perhaps the most aggravating and jarring, which renders this statue the most egregious candidate for formal demolition or casual destruction, and makes it a frequent target of vandalism and protest, for threats of dynamite and smears of the blackest ink, is that here in this most remote corner of Scotland, the inhabitants are daily compelled to see from all around them, and perched like a lightning conductor on the top of a noble Scottish hill, a statue to, of all things, an *Englishman*.

He was named George Granville Leveson-Gower, he was born in 1758, and he made his fortune—he was

The giant statue of the much-reviled first Duke of Sutherland has stood for 180 years on a mountain near the town of Golspie in northeast Scotland, its sandstone construction stoutly defying numerous attempts to destroy it.

so "abominably rich" that he was popularly described as "the Leviathan of Wealth"—from shrewd investments in the Industrial Revolution, compounded with an exceedingly astute marriage. During the seventy-five years of his quite extraordinary life, he carried a number of different names, displaying his steady ascent through the ranks of the nobility: he started life in Mayfair as the Viscount Trentham (his grandfather was named Scroop Egerton, the first Duke of Bridge-water); he became a member of Parliament under the

name Earl Gower; then became titularly demoted
to Baron Gower of Stittenham. Not content with his
personal nomenclature, in 1803 he became the second
Marquess of Stafford before finally, in 1833, becoming
tagged with the name by which he is best remembered:
the first Duke of Sutherland.

But his possession of this title was rather more brief
than the ill repute that would follow him. He ascended
to the dukedom on January 14. He dropped dead in
Dunrobin Castle on July 19. The 186 days of his title-
age was to be followed by an eternity of repudiation.
Blow up his statue, many now say. Scatter its fragments
like the ruins of Ozymandias, to lie forever among the
heather, so that for generations to come his huge sandstone nose will rise from the ground, his ears, as big as
horse troughs, will provide bathing for seabirds, and
families can take their ease and stage picnics on his
long, gartered legs.

For George Granville Leveson-Gower, the Duke
of Sutherland, was a sworn enemy of the Scots people,
they say, every bit as evil a despot as Hitler or Stalin. The memorial to his passing should be the long
memory of the misery he created for thousands, and
not a nobly carved behemoth of red sandstone, sculpted
for all the world to see.

Between 1807 and 1821, agents working for the

Countess of Sutherland* and Lord Stafford, as he was then still known, forcibly and cruelly removed thousands of crofters from their pitiful smallholdings and settled them, mulish and unwilling, scores of miles away from home. Of this there is no doubt. Nor is there any serious dissent as to the proximate reason. The removal—the clearance—of such people from the lands they had tenanted and tilled for generations in the glens and straths of the Scottish Highlands was a wholly economic matter. It was realized that the Sutherland estate, which at the time was the largest privately held estate in Europe, could and would be improved to rake in greater profits if the previously crofted lands could be given over entirely to *the raising of sheep.* Mutton and wool were of far greater value than the rents that might be charged to those who ran crofts.

And the ultimate reason for this realization was the Industrial Revolution. Steam-powered factories were opening in their scores, most at the time concentrated in the fast-growing cities of central England. The Sutherlands were prolifically invested in this unstop-

---

* She initially held the Sutherland title in her own right, having inherited the immense estates when she was just one year old. When her eventual husband became the second Duke, she became technically the Duchess-Countess of Sutherland.

pable new movement—and not least in the many new steam-powered factories that manufactured textiles. These textile mills, now busily and noisily creating endless yards of material for clothing men and women all over the world, needed wool, to be drawn from hundreds of tons of fleeces, in order to manufacture the more durable of their cloths.

The mill towns huddled around their factories also attracted what would be dense concentrations of people, all of whom now made money in the mills, and who needed to be fed and who earned the wherewithal to buy the necessary food. So sheep were to be the key. Sheep could satisfy the new needs of a new industry. The Sutherlands were canny enough to understand this—however brutally they behaved in deciding how best to meet those needs. The sheep would provide mutton for the people. And sheep, most crucially of all, would provide the fleeces, the ultimate source of wool, for the mills. In time, though none knew it at the point, steam railway trains would speed both meat and wool down to these mills, and further consolidate England's need for the goods and services of what would necessarily now become a much changed rural Scotland.

The story of how rural Scotland was so profoundly changed by this development has ensured that the

Highland clearances and the role in them of the despised Sutherlands have become both the stuff of legend and the victim of a quite singular degree of hyperbole. The most common and unimaginative comparison is with the Holocaust. "The victims . . . were subjects of intense hatred such as the Jews and the gypsies were to experience under the Nazis," wrote one F. G. Thomson in 1974. "Sutherland's managers kept records of their shipments of people," wrote a David Craig in 1990 "with the obsessional thoroughness of an Eichmann." And, introducing here the most notorious executor of the clearances, the Sutherland's chief factor, Patrick Sellar, an author named Eric Richards noted the existence of a book claiming that "Sellar's crimes against the people of Strathnaver were to be ranked with those of Heydrich, the man who perpetrated unspeakable acts against Jews in Prague in the Second World War."

Certainly the duchess herself was not much taken with her tenants. In 1808, at the time she and her husband were rolling out their plans for the improvement of their holdings' rickety financial condition, she noted that her lands were "an object of curiosity at present, from being quite a wild quarter inhabited by an infinite multitude roaming at large in the old way, despising all barriers and regulations, and firmly believing in witchcraft."

Her sentiments were much magnified by Patrick Sellar, the punctilious and unyielding Enlightenment-era lawyer whom she employed in 1811 as her factor—a figure who, in Lady Sutherland's particular case was, essentially, her consigliere. He carried out orders without mercy. He referred to the subsistence crofters of Sutherland as *primitives* and *aborigines* and felt no moral scruple in turfing them off the land as mercilessly as he could, all in the name of improvement.

The Sutherlands were not the architects of the Highland clearances—removals of uneconomic tenant farmers had been going on elsewhere, in the Scottish Lowlands too, at the end of the eighteenth century. Nor were they the practice's principal practitioners. But because of the particular and pitiless nature of their removals, and most especially those conducted by Sellar, theirs is the name that has become a byword for the brutality of the period, an ineradicable aspect of history that has driven a wedge between Scots and English for two centuries and more.

So many clearances were carried out under Sellar's direction that all—perpetrators and victims—became habituated, with a familiar unfolding pattern, a mixture of formality, fierce protest, and then disgruntled acquiescence. First, Sellar and William Young, a colleague from southern Scotland who was his agricultural expert, together with a team of others, hard Highland henchmen in tweeds and waterproofs, would set out from their headquarters among the fantastical towers of the Sutherlands' vast Dunrobin Castle, near Golspie, and travel by stage or trap or simply mounted on ponies into the barren wilds of the estate's vast landscape. Sellar would arrive at a crofter's door, his hard Highland sentries gathered behind him, all muscle and menace. When the door was opened, he would present the head of the household with papers that, he would explain, were instructions from Lord Stafford formally curtailing the current tenancy. Moreover, the orders then gave notice of the necessity that the crofter and his family vacate the premises, made it clear that alternative accommodations had been made available for them nearby, and gave notice that the quitting and relocation should be done in thirty days or less, with no room for dissent, argument, appeal, or resistance.

Sellar would then doff his cap, offering his respects to the family, and he and his team would withdraw, board their various conveyances and ride on down the strath to the next croft in line and make the same announcement yet again, on and on across the length of the valley. The crofters, shocked and frightened, stood at their doors bewildered, in most cases and for a brief while not knowing what to do.

The Sutherlands were bent thereby on enforcing a program of social engineering seldom attempted before anywhere, by anyone. The huge estate was already busily constructing roads and bridges (half paid for by the government) and extending postal services and civilization, so called, into the region—yet having still to deal with the economic and social inconvenience of a native population who lived in a manner the estate managers considered backward and barbaric, and in a place into which the light of modernity was being so benevolently extended. This to the Sutherlands was an outrage that needed to be extirpated.

The plan that the estate owners had contrived called for the enormous tracts of lebensraum taken up by these supposedly barbarous people to be vacated in favor of large sheep farms; and the former inhabitants to be given instead smaller three-acre crofts along the east coast of Sutherland—the name of the county, naturally. As a generous bonus, a town called Helmsdale would be built specifically for them. From this new settlement, once they learned the craft, they could begin to practice fishing.

After all, the North Sea was known to be brimming with fishable treasures. What better way for these impoverished *indigenes* to prosper than to learn how to win for themselves the bounty of the seas? Was not her

The immense Italianate pile of Dunrobin Castle offers a stark contrast to the modest sod-roofed cottage of a kind inhabited by the great majority of the Sutherlands' crofting tenants.

ladyship wise and benevolent beyond measure in decreeing that such should improve the lot both of her people and of her lands?

And what lands they were! The romanticized view is that the county is today one of the last *untamed wildernesses* of the British Isles—endless stretches of wild desolation, single-track roads, glacier-smoothed plains pockmarked with lochs, and in the western distance huge and strange-looking peaks of ancient rocks, nee-dling between them long tongues of the ocean with villages nestled by the white sands at the beachhead. There are wide and grassy valleys, the more fertile straths of the county, where before the clearances began in earnest in 1807, most of the crofters had been settled since medieval times. Every so often in these straths today you will pass a neat white cottage, most probably owned by a Londoner who comes up in the summer only; there are lichen-colored stone walls, a few bare and wind-blasted trees, daffodils in springtime, blue-bells, sheep everywhere, red deer, eagles, crying gulls.

But it is perhaps better to describe the county less as a wilderness—a word that suggests the overwhelm-ing power of nature against which mankind has failed to struggle with success—as a landscape of *derelic-tion*, in which mankind was a dominant and in later times a malevolent force, and whose agents left a leg-acy of ruin and despair and choked the life from the place. And so much of Sutherland is a place of ruins today, broken old houses that litter the moors as mute testimony—abandoned villages, forgotten churches, cottages without roofs, tumbled piles of black stone, mossy graveyards seldom visited and with their rusting iron gates swinging in the gales.

True, there can be days of endless sunshine when

the Sutherland landscape may even look inviting; but more often, huge gray clouds boil up over the great crags of Suilven and Canisp and Stac Pollaidh and a miasma of sheeting rain and fog settles down over the miles of heather, and the place then looks blasted, melancholy, even menacing. On days like that, it is easy to imagine Patrick Sellar and his men advancing on a terrified Highland family gathered at the door of their croft and telling them to be off, and best he quick about it, on orders of the castle far away.

It is fitting to recall W. H. Auden's terrifying poem "O What Is That Sound," when he speaks of the posse of soldiers down in the valley, unexpectedly turning to come up to your very house, massing at the gate outside your croft and then

> *O it's broken the lock and splintered the door*
> *O it's the gate where they're turning, turning;*
> *Their boots are heavy on the floor*
> *And their eyes are burning.*

The former village of Kildonan is the place most keenly remembered today for the perceived outrages of Patrick Sellar and William Young. A series of violent evictions took place there in the summer of 1813 in the valley just a few miles inland from Helmsdale. They were violent because for the first time since the removals program had begun in earnest across in nearby Assynt six years earlier, the enforcers were met with resistance. Though news travels slowly in the Highlands, it is certain that by this time—six years!—the crofters of Kildonan had heard of the forty-eight entire villages that had been ravaged in Assynt alone, had heard of the manner in which the factors had set fire to the crofters' roofs, how they had driven away their cattle, had set the pastures ablaze, had forced whole families—pregnant women among them—out onto the moors at the end of blackthorn sticks and brandished broadswords.

They well knew by then that Sellar's modernization strategy was to leave the sacked villages entirely uninhabitable, such that only sheep could henceforward live on the land and that the huge envisioned sheep farms— many of them run by Englishmen, to add insult to local injury—could be firmly and swiftly established, providing even greater rental revenue for the Sutherland estate. So by the time Sellar arrived on his first scouting mission to Kildonan, the men of the settlement were waiting for him. They were all too well aware of the imminent danger that was facing them, were wanting to have nothing to do with the niggardly three-acre coastal plots that were being offered in recompense, were having no truck with what most saw as the madcap idea of having to learn how to fish—when most could not even swim. They were, to a man, bent on open revolt.

At first, Sellar and the agents from Dunrobin Castle proved no match for the angry Highlanders, crowds of them blocking the newly built road that ran alongside the languid reaches of the salmon-rich Helmsdale River. Barricades were erected. Stones were thrown. Sellar was unimpressed, though puzzled. He was a no-nonsense man, a zealot of implacable determination, a lawyer-farmer who had already himself embraced modernization on his own property. He believed that so far as the stewardship of the barren lands here was concerned, there simply was no economic future for tiny tenanted farms. He was not merely acting to inflate the riches of his employers; he was striving to improve the overall lot of those whom he was trying to persuade to leave. What he was attempting to do was, he felt, for the crofters' own good. Moreover, Sellar's own grandfather had been cleared from his smallholdings in central Scotland many years before and he, the grandson, had prospered nonetheless. No doubt these crofters of Sutherland, after the initial shock, would set about improving their own lives, would one day come to accept that Patrick Sellar had done much good for them and for Scotland.

But the men of Kildonan were having none of that. The barricades stayed put, the stones continued to fly, the ponies on which the eviction squad were traveling were harassed and frightened and repeatedly tumbled their riders into the heather. Each time the team returned to Dunrobin without having budged the villagers, a perplexed Lady Sutherland told them to try again the following day. She was shocked, shocked, at such hurtful rejection of plans that she believed were wise and benevolent.

Her patience finally snapped in August. Sellar and Young, she declared, needed brute force augmentation. Using her influence with the local colonelcy, she was able to dragoon a local unit of the army into standing alongside her eviction squad. With bayonets at the ready, the platoons of infantrymen, terrifying to behold and impossible to resist, eventually forced the villagers to stand aside.

Once the houses had been cleared, Sellar did what he always did: he briefly allowed the villagers time to harness their livestock and drive them onto the Helmsdale road, and then set the heather ablaze to spoil the pasture, and he had his men break the roof timbers and set fire to any thatch, thus ensuring the entire village of Kildonan was wrecked and useless. It was My Lai, Scotland-style. Kildonan's ruins are visible today, just low walls of tumbled stone poking vaguely up through the grasses, outlines of lives that were moved elsewhere, like it or not. There is a small Church of Scotland, most of it built in the 1780s, now almost abandoned, but with plaques memorializing the fate of the evicted highlanders.

And for the evacuees, lives changed greatly. They realized as they had suspected that the small and pinched crofts in and around the harbor at Helmsdale were unsuitable and unacceptable. After a lengthy conclave it was decided that a number of the evacuees would abandon not just Sutherland, but Scotland too—and would flee instead to Canada, and to the widely advertised Red River of Manitoba. They had heard, and rightly so, that the lands around this faraway river were owned by a Scotsman, who would look kindly on them settling there.

Accordingly they boarded the SS *Prince of Wales*, a Hudson's Bay Company freighter that was already on migrant passage from Stromness in Orkney clear across the North Atlantic to York Factory, a company fur-trading settlement on the western shore of Hudson Bay. The vessel's surviving passenger list shows a total of ninety-six passengers. While a few of them were collected from a brief stop in Ireland, most—names like John Sutherland and family, Jean and Angus Mackay,

Betty Gray, the widow Barbra McBeath—were names from the old parish rolls of the Strath of Kildonan. They left Helmsdale in August; they had two terrible months on the little vessel—typhoid killed many; seasickness carried off others—and the much diminished party finally arrived at the tiny factory in October, just as the winter snows were settling and the rivers were icing over.

A small number of more intrepid evacuees,* all kitted out with company-supplied snowshoes, then made their way southwest toward what was called the Red River Settlement—trekking across the tundra, eventually camping on some 116,000 acres of land that the Hudson's Bay Company just months before had sold, and just as the migrants had heard back home, to a fellow Scotsman, the Earl of Selkirk, who entertained the hope of establishing a homeland for dispossessed Scots like these. Their weeks-long expedition was fraught with difficulty—not least because they walked unwittingly into all manner of disputes between trappers and traders and rival claimants to the lands and the trad-

---

* One of those Kildonan exiles listed on the manifest was a George Bannerman, whose great-grandson would be John Diefenbaker, Canada's amiable prime minister for six years beginning in the late 1950s.

ing opportunities on the flatlands around the river junctions ahead. But settled they eventually did, and in time their settlement became the core of the city of Winnipeg, capital of the Canadian province of Manitoba. From time to time citizens of Scots ancestry make their way back to Kildonan, to pay homage to their melancholy beginnings.

There is an inescapable irony to the crofters' settlement in Canada. The land on which they would build the foundations of the city of Winnipeg, and of which history all too casually records the ownership of Lord Selkirk, did not of course "belong"—again, the concept of land "belonging" is not universally recognized—to any white man. The acreage had for years been the homeland of métis people—descendants of unions between French fur trappers and local Indians—who had squatted there, admittedly without title, and who were understandably hostile to the newcomers. There were fights and killings and on occasion soldiers had to be brought in to separate the warring parties. And other Northern Plains Indians—such as the birchbark-canoe builders, the Ojibwe, who are still so prominent in the headwaters of the Mississippi River, some short distance to the south over the American frontier—also declared an interest in the land. Though all is peaceful now for the Scots in Manitoba, there must have been times when the Strath of Kildonan seemed a relatively serene place, in spite of the menace of Patrick Sellar and his gangs.

Sellar himself lived on for more than three further decades after the height of his clearance activities, dying in 1851, unloved and widely cursed (such that flowers never grew on his grave, it was said) but with a professional reputation as an intelligent farmer and prescient reformer somewhat enhanced. Had he not existed, then maybe nor would Winnipeg.*

Meanwhile, Scots of the more radical nature are content to continue their assault on the memory of Sellar's employers, the Duke and Duchess of Sutherland. And to do all they can to hasten the removal of the hated "Mannie" from on top of the summit ridge of Ben Bhraggie, in which ambition it has to be assumed that one day they will succeed.

---

* Nor would one of the more perennially amusing books of the twentieth century, the riotously successful historical parody *1066 and All That*, published in 1930 and never since out of print. Its authors were R. J. Yeatman and W. C. Sellar, the latter a schoolmaster who was also the grandson of the notorious Patrick. The contribution made by Walter Carruthers Sellar by way of this remarkable and still-selling book amply compensates, in some English eyes, for the cruel and brutish behavior of his reprehensible grandfather.

2

The Accumulators of Space

Boone Pickens' Mesa Vista Ranch:
Roberts County, Texas.
64,809± Acres
An Oasis in the Texas Panhandle—
the World's Best Quail Hunting.
Price Reduced to $220,000,000.
Chas. S. Middleton and Son, Lubbock, Texas.
—Advertisement in *The Land Report* (Winter 2019)

The biggest personal landowners in the world are nearly all monarchs or absolute rulers of one kind or another: the kings of Swaziland, Bhutan, Jordan, Nepal, Thailand, Morocco, Saudi Arabia, and Lesotho, the Emir of Kuwait, the Sultan of Oman, the Pope, and the Sheikh of Qatar. The list necessarily also includes Britain's Queen Elizabeth II, who is technical owner of last resort of the entire acreage of the United Kingdom, from Shetland to the Scilly Isles, together with portions of—in most cases, all of—some fifty-four now independent nations that once made up the British empire, with her still being the titular queen of thirty-two of them. A quarter of the world's population lives on land in which, though the individual citizens may not know it, they exist in a notionally feudal relationship with the British Crown.

In strict technical and legal terms this British monarch exercises stewardship over the virtual entirety of Australia, Canada, and New Zealand, realms that together make up a sizable portion of the Crown's 6600 million acres. In many other democratic countries—Germany, France, and the United States, most obviously—an individual may own land subject only to a few restrictions. Where the relict British writ still runs, the arrangement is more akin to vassalage, if much diluted and with any slight historic obligation seldom if ever enforced. The somewhat-past-its-sell-by-date theory holds that if the true and ultimate owner of all the Earth is God, and if the current holder of the British Crown is God's representative on Earth, then it surely follows that the monarch holds the land in trust on God's behalf, and for the Deity's convenience and pleasure. Few but the most worshipfully doctrinaire seem not to chafe at the absurdity of this arrangement.

Of course, when considering how owners steward their lands it is hardly possible to judge those who own entire countries. Queen Elizabeth can hardly be expected to know how well or ill a wheat field in Saskatchewan is being looked after; nor can the King of Saudi Arabia, a man with a myriad pressing concerns, be too bothered about the condition of every dune in the Rub' al Khali. The granularity of stewardship can really be considered only by looking at the landholdings of those lesser individuals, men and women who own great tracts yet are by no means sovereign in doing so.

Such landholdings by private individuals in Britain—difficult to ascertain in detail, so notoriously secretive in such matters is the still class-bound and suspicious country—are sizable, though they are understandably modest in comparison with the vast tracts owned privately across the huge fastnesses of the United States and Australia. The largest personal landholding in Britain is probably that of the Duke of Buccleuch, who is listed as presiding over some 270,000 acres, though at the time of this writing he's reported to be selling off fields and farms at a dizzying rate. All told, the British dukes own a little more than a million acres of British land, which, though plenty, is as nothing when compared with the world's major possessors of territory.

By most accounts, the world's very largest private landowners are all Australian—which is hardly surprising, given the size of even quite ordinary cattle or sheep stations. The largest single property in the country, the Anna Creek sheep station in South Australia, is almost 6 million acres in extent—which, when one considers that all of England is just 32 million acres, puts English landowners very much in their place. Indeed, when they are combined, the ten largest stations in Australia constitute rather more in land area than exists in all of what was formerly the mother country.

One Australian owner looms large over all others, however. Gina Rinehart, heir to one of the country's largest iron-ore-mining fortunes and no mean businesswoman herself, appears currently to be the largest private landowner in the world, with 29 million acres under her various companies' ownership and control. Others were jockeying for the title, one of whom, the Kidman family—the actress Nicole is a relation—was a close enough rival that Ms. Rinehart, brooking no competition, bought them out to assure her primacy. Unlike the only marginally larger England, much of her land is underpinned by deposits of iron, and with

China a ready customer just a few days bulk-carrier-sailing-time to the north, her companies make enough money to keep her at the top of the Australia's Richest Person and World's Richest Person and World's Richest Woman lists, hovering in the single digits in most of them. She enjoys a flotilla of large houses, engages in keenly fought legal disputes (most especially with her much younger Filipina stepmother in a money-related case that lasted for fourteen years and kept the readership of Australia's colorful tabloid press highly enthralled). She is also one of the rare individuals to own an upper-deck three-bedroom apartment on the perpetually-globe-circling ultraluxury vessel MS *The World*, a ship rather cattily described as having been built for those who are wealthy, but not quite wealthy enough to have their own cruising yachts.*

With Ms. Rinehart's properties being so extensive, she might as well be head of a small sovereign nation. Much the same difficulty exists in trying to assess how her lands are run as in trying to assess the environmental impact of the King of Bhutan's ownership of Bhutan or the Pope's effect on the Vatican's extensive holdings in Argentina or Portugal. All that can reasonably be

---

* The American television personality Judy Sheindlin—known on-screen as Judge Judy—is another of the ship's residents.

The gigantic inherited landholdings of the wealthy mining heiress Gina Rinehart have inspired more awe than affection among her native Australians, although colorful court battles involving family members have done much to humanize her image.

ascertained derives from her publicly expressed attitudes. It is known, for example, that Ms. Rinehart contributes significant sums to organizations that promote climate science skepticism. She has written widely on the subject, declaring in 2011 that "I have never met a geologist or leading scientist who believes adding more carbon dioxide to the atmosphere will have any significant effect on climate change, especially not from a relatively small country like Australia."

She also seems rather less than amiably disposed to the broad spectrum of working-class Australians, suggesting that they should stop "whingeing" and should "do something to make more money yourselves— spend less time drinking or smoking and socializing, and more time working." Warming to this theme in one of her rare interviews, she went on to suggest that if Australians "competed in the Olympic Games as sluggishly as we compete economically, there would be an outcry. Africans want to work and its workers are willing to work for less than two dollars a day. Such statistics make me worry for this country's future."

Gina Rinehart's precise views on Australia's 800,000 aboriginal people, and on the land to which they would seem quite as morally entitled to live and regard as their own as are the Native Americans in the United States, are little known, though doubtless her companies employ large numbers. If her views are uncertain, those of her notoriously bluntly spoken father, Lang Hancock, are often quoted in Australia. He had little time for the indigenous people, it would seem, nor for their claims to their land which he saw as rich with iron ore, and needing to be mined, not preserved.

"Mining in Australia," he once said, "occupies less than one-fifth of one percent of the total surface of our continent and yet it supports 14 million people. Nothing should be sacred from mining, whether it's your ground, my ground, the blackfellow's ground or anybody else's. So the question of Aboriginal land rights and things of this nature shouldn't exist."

He was still more forthright in expressing his views in 1984, when he went on television to suggest forcing unemployed indigenous Australians—specifically, "the ones that are no good to themselves and who can't accept things, the half-castes"—to collect their welfare checks from a central location. The reason for this became clear when he further explained that "when they had gravitated there, I would dope the water up so that they were sterile and would breed themselves out in the future, and that would solve the problem."

The notion that the sins of the father might well have been visited on his similarly blunt and forthright only daughter is perhaps not wholly unimaginable. Meanwhile, her attitude to the land she possesses is defined entirely by the profit she can wring out of its mineral wealth underground. Money, most Australians say of her, remains her greatest obsession, and with China, Japan, and Korea to her north eager for every scrap of hematite they can get their hands on, and with her companies owning 29 million acres of invariably iron-rich Australian landscape, the ultimate fate of her lands has to be predictable: for a while at least, they will be quarried and the ore gouged out of them, and then they will be returned to something like their original state in the way that so many mining companies around the world are legally obliged to do.

And as for what Australian mineral magnates like to call the T.O.s, the Traditional Owners? Lang Hancock had no time for them, and for a while paid such aboriginal workers as worked for him in food and tobacco, rather than money, which he believed they would only spend on drink. He thought—as did most early Australian settlers—that most of the Australian interior was *terra nullius*, land that belonged to no one and so ripe for such exploitation as he thought fit.

The law has changed in recent times, and with it the recognition that much of what was thought *terra nullius* has in fact been occupied and owned, as it were, by what is generally now accepted to be the oldest continuous civilization on the planet, some sixty thousand years old—far older than the Egyptians, the Chinese, the Greeks. The Australian government recognized such a reality too, and now requires all dealings in what once was called the *gaba*, the Great Australian Bugger-All, to take exceptional care of the traditional owners' history, needs, and people, with compensatory payments and respect offered in equal measure. And to be fair to Gina Rinehart, she seems to be following these rules, and there is no evidence that she is reluctant in doing so. She wants her iron, she wants her money—and if there are rules that inhibit her riding as roughshod over the Australian landscape as once her father did, she is complying and not complaining.

There are a fair number of mightily wealthy landowners in the United States, of course. The twenty biggest own well over half a million acres apiece, and together the top hundred own as much land as the entire state of Florida. And the rate of expansion of private ownership is phenomenal: since 2007 the amount of American land owned by these wealthy one hundred has increased by 50 percent and is showing no signs of slowing down. Many of them are ranchers, and invariably own tracts in Texas and in the cattle-friendly western states. A great number of their holdings go back to forefathers who ran cattle in the nineteenth century, and these landowners attach some reverence to their territories, seeing responsible ranching of the western emptiness as something of a public trust. Other large holdings have been acquired by those who believe that lumber—for home building or papermaking—is the more sustainable and perhaps easier way to make a fortune; their lands tend to be in the wetter, hillier territories, either in the Pacific northwest or the forested expanses of the northeast.

The names of these men and women who have dominated the North American *corps de propriétaires* for decades are not as nationally familiar as some—there are the Stimsons with their forests in Oregon; the Martins' lumber operations in Louisiana; the O'Connors's Texas ranches near San Antonio; the Hamers in West Virginia; the Lykes family (owners also of the famed Lykes Brothers Steamship Company), with their once lucrative trade of lumber between Florida and Cuba; the coequally gigantic Briscoe and King cattle ranches, both in south Texas; the Pingrees, who were nineteenth-century slave-traders, now with their massive holdings in New England; the Irvings in Maine and New Brunswick; the Reeds and their timber operations in Oregon; and the Emmerson family, with their reputedly well-managed forests in the high inland mountain ranges of California. None of these clans is so famous as their wealth suggests, and they are little known beyond their immediate neighborhoods.

Each of these families has a historic claim to notoriety. The descendants of the onetime Canadian sawmill operator K. C. Irving, for example, now have a near total hold over the small New Brunswick city of St. John, running the power generation, the fuel supply, the newspapers, the docks, and many of the stores; in addition, despite a colorful series of internecine squabbles of epic proportions, the principal family company, J. D. Irving Limited, also runs a range of businesses that includes paper and pulp, railroads, trucking, shipbuilding, chandleries, frozen potatoes, packaging, crane rentals, plastics, an ice hockey team, twelve newspapers, a shipping line, a bus company, and a tire distributorship. All of this, together with 1,247,880 acres of forest land in New England and eastern Canada—on which the family claims it plants 20 million new trees each year, demonstrating their reputed commitment to the sustainable management of their immense territorial resource.

These are the old, somewhat conventional, traditional landowners. The very largest of all nowadays, however, seem to be newcomers—men in the main, vastly successful in their various fields, which have little to do with land itself, but are usually but not invariably associated with either the electronic arts or with fast food, and who decided in later life to invest in the one immutable and incorruptible investment that will seldom be much affected by any turmoil in the market. They seemingly felt a compulsion to own very large tracts of American land, and to employ it in a variety of ways.

Ted Turner, for example. He founded the Cable News Network, CNN, in 1978, pledging his then already considerable fortune to keep gathering and broadcasting television news "until the world ends." He started many other ventures—wrestling, old movies, Russian TV, gift giving—managed to get himself embroiled in any number of vicious spats with rivals in an increasingly competitive business, and then began buying land, in immense amounts. He was by birth an Ohioan, by persuasion a southerner, but in terms of environmental activism became, like so many romantics, a devotee of the American west. So the land he bought was largely in Montana, Kansas, Oklahoma, Nebraska, and New Mexico—1.9 million acres in all, with the largest single parcel being a ranch in far northern New Mexico, on the Colorado state line, with almost 600,000 acres of semi-arid desert land, half of it in the Great Plains, the western half in the Sangre de Cristo Mountains.

The history of this single parcel is probably unique in the history of the American west—a former tribal home to the Apache Indians, it was the centerpiece of a huge land grant* once made by the Mexican govern-

---

* The U.S. government has made much use of land grants—whereby tracts of publicly owned land (originally taken, of course, from Native Americans) are given away to allow the financing of projects deemed for the common good. The three Morrill Acts (the first two signed by President Lincoln) allowed 17 million acres of federal land to be sold to allow for the establishment of public universities like the University of Massachusetts, Rutgers in New Jersey, and Cornell in New York. Since the

The founder of CNN, Ted Turner, was until recently the owner of the greatest acreage of private land in the United States until a rival TV executive, John Malone, outdid him. Turner has done much to encourage the restoration of the bison population in the West.

ment in the 1840s to a French fur trapper, whose son-in-law, a celebrated Irish French mountain man named Lucien Bonaparte Maxwell, then found gold in its mountains, expanded his holdings, sold them en masse

grant for Cornell was thought insufficient, the institution was allowed to pick unused public land elsewhere; it chose woodlands in Wisconsin and made enough money from the sale to endow one of the country's greatest centers of learning.

to a British company. After this (and after the killing of Billy the Kid on one of the ranches in 1881) the grant was eventually broken up into a raft of smaller—but still enormous—properties, one of them owned today by the National Rifle Association, another by the Boy Scouts of America,* and the largest parcel of all, known now as the Vermejo Park Ranch, by Robert Edward Turner III, of CNN.

Ted Turner, whose trenchantly expressed political opinions are unconventional among the American elite—he believes, for example, that Iran should be allowed to have nuclear weapons, that the Palestinians are no more inclined to terrorism than is the state of Israel, that there is little brutality within North Korea, that women have an absolute right to abortion and that organized religion is generally to be deplored—is a committed environmentalist and a true believer in the perils of man-made global warming. He spends millions to protect endangered species—most recently the reintroduction of the Bolson tortoise, *Gopherus flavomarginatus*, on one of his New Mexico properties a couple of hundred miles south of Vermejo Park. This

---

* The world's only paleontologically confirmed track of the footprints of a *Tyrannosaurus rex* were found on the Scouts' land here in 1993.

act of biological munificence won him accolades from the Turtle Conservancy, who now declare this giant and once endangered species to be flourishing happily again. And enlightened self-interest plays a part too: Turner owns a chain of forty-odd restaurants, Ted's Montana Grill, and is doing his best to persuade the American dining public to consume cuts of that most emblematic of American ungulates, the Plains bison. He has made considerable efforts in recent years to reintroduce bison to his High Plains holdings, harvesting them responsibly, he claims, and managing their populations in a common-sense manner.

Reintroducing the bison to the American west has become something of a mission for Turner, an otherwise agnostic figure whose current motto is *Save Everything*. He claims to be only too well aware that the appalling behavior of nineteenth-century settlers brought these magnificent mammals to the brink of extinction. The argument at the time was more anti-Indian than anti-buffalo: those promoting western settlement held that by denying the hunting parties of Plains Indians their own communities' main source of meat, the number of Indians themselves would be reduced by starvation, making the land easier to claim by the incomers. It was not long before this ideologically justified reason turned into an unstoppable orgy of bloodlust, and in their thousands gangs of huntsmen were soon thronging the prairies. Most came armed with large .50 caliber weapons, which aficionados considered the best means of bringing down the gentle, placid, lumbering, multiton grazing creatures.

Hunters would come to hunt them by train, for example, as *Harper's* magazine once reported:

> Nearly every railroad train which leaves or arrives at Fort Hays on the Kansas Pacific Railroad has its race with these herds of buffalo; and a most interesting and exciting scene is the result. The train is "slowed" to a rate of speed about equal to that of the herd; the passengers get out fire-arms which are provided for the defense of the train against the Indians, and open from the windows and platforms of the cars a fire that resembles a brisk skirmish. Frequently a young bull will turn at bay for a moment. His exhibition of courage is generally his death-warrant, for the whole fire of the train is turned upon him, either killing him or some member of the herd in his immediate vicinity.

Fifty million bison were slaughtered by such fish-in-a-barrel methods during the mid-nineteenth century. Indians would sit on their horses and watch, openmouthed with horror at the mob behavior, as trains would pass by with hundreds of gun barrels firing endless cannonades, as if battleship broadsides. And once the train had passed over the horizon, and the wind-soothed silence returned to the prairie, so these same Indians would gaze uncomprehending at the sight of hundreds of bison carcasses lying beside the tracks, soon to be bloating and rotting in the scorching sun. The hunters never even bothered to collect their kill, being interested only in the act of killing itself. Some entrained visitors claimed to be bagging six thousand animals a season, and popular images were published showing Matterhorns of discarded bison skulls waiting to be mulched into fertilizer.

Turner remains appalled by what happened. His efforts to breathe life back into the bison population have been rewarded: he now raises some fifty thousand animals himself, and with other breeders joining the buffalo bandwagon, the animal is no longer on the verge of vanishing. And while bison meat remains a niche product for the still beef-obsessed legions of American carnivores—the restaurant chain is just a moderate success—it can fairly be said that Turner's nearly two million acres have helped restore some corners of the Great Plains to something approaching their original,

It was briefly considered sporting fun to board a train in
Nebraska and head west over the prairies, blazing away with
rifles at any passing bison—in the aftermath, millions died,
and skulls were piled fifty feet high. Indians, who thought the
beasts sacred, wept, incredulous.

natural state, before the depredations of nineteenth-
century settlement brought ruin elsewhere.

Ted Turner is not the largest American landowner
anymore, however. That distinction goes to another
practitioner of the electronic arts, John Malone, an engineering billionaire born in Connecticut of Irish Catholic stock who also made his fortune in the early days
of cable television. He currently runs Liberty Media
and he owns the Atlanta Braves baseball team, a large

chunk of Expedia travel agency, the QVC shopping channel, and the Sirius satellite radio operation. He is, in other words, a major player in the twenty-first-century American economy. He also owns an impressive 2.2 million acres of the land surface of the United States—and 3200 more acres in Ireland, where he has a large fairyland castle in County Wicklow as well as one of the finest Georgian mansions in the country, Castlemartin, on the banks of the Liffey in County Kildare.

In 2011, after years of chest-bumping competition with Ted Turner, Malone won the race to become the owner of the most land in America. He began his acquisitions in the West, with a million or so acres gathered up in and around the Rocky Mountains in New Mexico, Colorado, Wyoming, Montana,* and Nebraska.

---

* Montana—Big Sky Country—has become the most magnetic of temptations for landowners big and small. I once fell under the state's spell and, with a wilderness fantasy born of Thoreau and Norman (*A River Runs Through It*) Maclean, bought eighty acres of open land on the east bank of the Bitterroot River for $40,000. My neighbors, had I ever been able to afford to build, were Huey Lewis, Christopher Lloyd, and Andie MacDowell. A year later, strapped for cash, I sold the land for a welcome $80,000. Ten years later, I returned and visited the parcel once again: the real estate agent was hesitant to tell me its most recent price: $1.25 million. I wept.

On portions of these tracts he raises and trains quarter horses and he breeds beef cattle—in huge numbers in Nebraska particularly, where his Silver Spur Ranches run feedlots on the banks of the North Platte River. More recently, he developed an interest in the American Northeast, buying tracts in New Hampshire and, in the effort that finally pushed him over the line in the Turner race, purchasing fully 980,000 acres of the far northern forests of Maine.

Once Malone had made the pole position he spoke, as he seldom does, of the ownership-addiction "virus," as he put it, that he supposedly caught from Turner. He repeats the well-worn dictum about one "never really owning land" and that "you are a kind of steward of it, so you really want to be a good steward. But it sure is enjoyable while you are on the planet to be able to go walk around or get on a horse and drive a truck around and be out in the open."

So far as is known, no particular animal or plant lays claim to John Malone as either its savior or guardian—no tortoise, no bison. He is an eager horse breeder and owner of thoroughbreds, both in Florida and at studs dotted around southern Ireland—and his cattle ranches in the west are said to offer "sustainable grazing practices to ensure the health and balance of native grasses and prairie, while keeping the carbon sequestered in undisturbed soils. Only horses' hooves and cow tracks impact grasslands and soils, storing carbon, reducing erosion and ensuring food security."

Of his expanses in Hancock County, Maine, the public relations agency that writes such material as this could say only that Mr. Malone "uses photosynthesis" to offset carbon emissions—omitting to tell the less sophisticated reader, perhaps, that trees, with which Mr. Malone's lands in Maine are entirely draped, perform photosynthesis without any urging from their owner. This, put simply, is what trees do. And the Hancock County land is all fir and spruce, maple and birch; there are moose, bears, lynx; there are streams choked with salmon, lakes with trout and bass. Basically, all of inland Maine is very much as an airline passenger might see upper New England unspooling below, all the way from the St. Lawrence River to the New Hampshire border: endless stretches of dark green forest, pristine and unused, wild and forgotten.

Within the state, though, there remains a stubbornly perceptible degree of concern as to what Mr. Malone might do with his immense acreage. Nationally, the Nature Conservancy has praised his efforts at improving his lands and preserving wildlife. But locally, they are not so sure. The Natural Resources Council of Maine well understands that there are logging operations in their forests generally—it would be idle to suppose otherwise: 95 percent of Maine's land is privately owned, and such inland industry as exists consists of lumber and paper and little else. But the council nonetheless presses new owners to put at least some of their holdings into conservation easements, so there can never be any environmentally ruinous developments—no nuclear power stations, no airfields, no getaway resorts. And it bothers them that thus far Mr. Malone has not signaled his intentions to do that—though he will continue the informal tradition in Maine of allowing public access, so that snowmobilers and skiers and those other harder breeds of men and women who like the cold wilderness of the far northeast may indulge in their pursuits, or gentler types may experiment with the forest bathing that is currently said to be so soothing for the soul.

America's two biggest owners do seem, in short, to feel a fair degree of responsibility for their immense holdings of the national landscape. They are seen as having little in common with the robber barons of old, men who in all too many instances scourged their territories and took from the land what they wanted, at will and with little regard for the future. One hopes and supposes the lands of such new masters as these will flourish, as will the animals and plants together, under their supervision.

Not that the robber barons have entirely gone away. Jeff Bezos, the founder and owner of Amazon, uses his 400,000 acres in far western Texas as a launch site for his Blue Origin space rockets. And the press-shy "Silent Stan" Kroenke, who owns sports teams—Britain's Arsenal football club among them—and stadiums, and who married into the immense Walmart fortune, recently bought the half-million-acre Waggoner Ranch in far northern Texas—said to be the largest ranch in the country that is surrounded by a single continuous fence. Soon after buying it, he sent eviction letters to a group of elderly and impoverished people who had lived for decades around one of his many lakes. The eventual cavalcade of ragged and impoverished old-timers—whose families before them had lived on the shores of Lake Diversion for decades under a carelessly informal arrangement with the previous ranch owners—presented a sorry sight; their loathing for Mr. Kroenke knew no limits, and they cursed him loudly as they passed the lake for the final time.

And then there are the Wilks brothers.

The Land Report, the Dallas-based quarterly bible for America's landed set, takes a decidedly sniffy tone in noting that the conservative evangelicals Farris and Dan Wilks, with their 705,475 acres and climbing—they acquired a further 3108 acres in 2019, and show no signs of halting their shopping spree—are currently the twelfth largest owners in the country. "They love land" is the only comment within the brief entry for the journal's annual top-100 list; one has the distinct feeling that the magazine's editors, who make fulsome remarks about almost everyone else on the list, do not much care for the Wilks brothers, for whom they have not a single kind word.

Popular disdain for the brothers seems to extend to those states—the mountains of southwest Idaho, principally—where they own most of their land. There and in Montana and down the Rockies all the way south to west Texas, the pair have in recent years created and expanded an enormous forest- and rangeland empire. In doing so, they have been busily forbidding all uninvited others from having access to their acres, no matter how lenient and permissive the previous owners might have been. They have hired armed security guards, thrown up gates, dug antivehicle ditches, strung barbed-wire fences across public roads, all intended to stop people from entering or crossing or even passing unacceptably close to the lands they now own.

"Many local residents," remarked a newspaper in 2019, "now see these new owners as a threat to a way of life beloved for its easy access to the outdoors, and complain that property they once saw as public is being

Forest roads like this, in Placerville, Idaho, were long used freely by walkers, snowmobilers, and the like—until the publicity-shy Wilks brothers, fracking billionaires from Texas, bought the lands and closed off all access.

taken away from them." And there have been street demonstrations, hundreds of people outside the Idaho government buildings in Boise—large crowds by Idaho standards—demanding that access to their forests and rivers not be interrupted.

The Wilks brothers began their own lives modestly, in the dry cattle countryside of Cisco, Texas, a hundred miles west of Fort Worth. Their father was a bricklayer, then ran a successful masonry firm. Once his sons reached their majority, they had a stroke of luck: they managed to get in on the ground floor of the fracking industry, setting up a company selling equipment that is used to pump chemically infused water at high pressure deep into oil-bearing shales, breaking up the rocks and allowing the gas and oil they contain to escape.

The business in which they played a critical part from its very beginnings is environmentally disastrous, but in the United States wildly popular—and similarly so in those countries that allow extractive industries like coal mining, oil drilling, and fracking to do more or less as they please. And under a succession of Republican presidents, governors, and Texas county bosses, so the Wilkses' company, Frac Tech, made them millions, and very quickly. In 2011 they sold their young firm to a group of investors, led by the Singapore sovereign wealth fund, for $3.5 billion, of which they each kept $1.4 billion. They now had unimaginable wealth almost overnight, and after little enough thought they opted to spend as much of it as they could on buying grand parcels of western lands.

It has not been an unalloyed success. Battles between private landowners and custodians of the public lands are nothing new in the American west, for sure—nor is rivalry between adjacent private owners either, as any student of American cinema will know all too well. But nowadays, with an increasing number of extraordinarily wealthy individuals buying up expanses of hitherto quasi-common countryside and—an echo of eighteenth-century England—enclosing them with fences, locals see the closing off of a cherished part of rural life. Men and women who have camped, skied, hunted, walked, climbed, and canoed in some of the loveliest and most pristine corners of the American outdoors now find barriers, guards, attack dogs, unfriendly signs—and within, they see private jets, hotels, airstrips and all the myriad appurtenances of extreme wealth, spy cameras meeting up with solitude, and neither much caring for the other.

And so the Wilks brothers have become symbolic scapegoats, villains in the growing confrontation between—as a columnist for a Boise newspaper had it—"Two American dreams." On the one hand all agree that there is an absolute right of private ownership; yet most surely also accept that there is a spiritual poetry to quiet pastoral beauty, and efforts to preserve such landscapes that offer such escape provide an incalculable benefit of the American public soul.

It does not help, does not mollify local sentiment, that the Wilks brothers say that in buying land and keeping it for themselves or selling it on to developers, they are merely doing God's will. They are intensely, and some might say eccentrically, religious. Their parents,

Voy and Myrtle, reportedly suffered the unusual punishment of being "disfellowshipped" from their initial
spiritual brotherhood, the theologically ultraconserva-
tive Church of Christ, apparently because they were
eager to cleave to an even stricter doctrine, even more
rigid dogma. The sons then went one step further than
their parents, establishing their own church, that of the
Assembly of Yahweh (7th Day), which has a bewildering set of rules—it adheres to traditional Jewish rites, its
adherents keep kosher, follow the literal interpretation
of the Old Testament, have Saturday as their Sabbath,
do not mark Christmas, Easter, or Good Friday, and so
far as public morality is concerned, regard abortion in
all circumstances as nothing less than murder and ho-
mosexuality as a grievous crime, a "base and demented"
practice that could see "the end of our nation" and the
"breaking of Yahweh's covenant," as the Wilks have
said in their sermons from the Assembly pulpit.

More concerning, so far as their ownership of land
is concerned, is the brothers' utter disdain for matters to do with environmentalism and with the fiction,
as they see it, that is climate change. "We didn't create
the earth, so how can we save it?" is Farris Wilks's oft-
repeated refrain. "When you realize that Yahweh is in
control it's much simpler—you can turn over some of
those responsibilities to him. And if the polar ice is get-

ting a little scorched, well, maybe that is just a message from God." Pedophilia and bestiality, the brothers believe, are soon likely to become legal in America. The nation is fast heading for the scripturally predicted end times, an armageddon, a rapture, brought about by humankind's eternally sinful and nonbelieving practices. They believe Donald Trump the best leader to push back against such evils, the man most prepared and determined to repair the damage already done; they have given millions to support him and those causes that are dear to his heart.

These are the people who currently own 705,000 of America's acres. In the name of their god, they have since put up their fences, have closed access to roads still considered by the state as belonging to the public at large, have turfed loggers off their lands, have forbidden snowmobiling, skiing, and snowshoeing. To ensure compliance they hired a lobbyist in Idaho with a view to changing the law on trespass, for "our Heavenly Father has blessed us with lots of gifts, and our family's priority is to protect them." And under the relentless pressure of these extremely wealthy men, a raft of enhanced antitrespassing measures has indeed now become settled Idaho law.

3

Going Nowhere and Everywhere

Cuius est solum, eius est usque ad coelum et ad inferos. [*]
Whoever owns the land owns it all the way to the heavens and to hell.

—Attributed to FRANCISCUS ACCURSIUS,
Italian jurist (13th century)

---

[*]   The invention of the airplane played havoc with this ancient maxim, and for years cases were brought suggesting that a flying machine crossing above your house constituted a trespass. A famous 1946 case, *United States v. Causby*, decreed an upper limit to owner exclusivity at 365 feet—one of the reasons that drone fliers nowadays, and out of an abundance of caution, fly their devices above 400 feet, still able to snoop, but in the legal free zone.

C entral to the concept of owning land is the right to tell others to get off it. One who acquires land gets to enjoy the legally famed Bundle of Rights: the right of possession, the right of control, the right of enjoyment, the right of disposition—and, most relevant here, the right of exclusion. A landowner may exclude others, may forbid others to stray onto his property, and has a right in law to demand that enforcement officers compel the person who does so—who *trespasses*—to leave.

That is the theory, a legal notion that is as old as the hills, almost literally. But in practice there is a vastly wide spectrum of attitudes toward such misbehavior. In all American states, for example, trespass is seen as a serious violation of personal space, and the trespasser's failure to leave when asked or told to is an offense. Elsewhere, at the polar opposite end of the spectrum, there are countries where it is entirely legal— subject to certain eminently reasonable restrictions— for anyone to be on privately owned land, whether invited or not, and it is an offense, albeit a minor one, to order them to go away.

Trespass—the word is French in origin, and signifies in its general, extralegal sense a transgression, and in a more general courtroom sense *a passage across the boundary of law*—is distinguished by three very specific legal senses, which are recognized terms: *trespass to person* (assault being one), *trespass to property* (damaging someone's possession an example), and the one that most concerns us here, *trespass to land*, which has been a legal concept under that specific name since the fifteenth century. It is a tort, insofar as it is a breach of the landowner's right to keep people off his land, and so it is a civil offense—it becomes a criminal offense only if damage is done, or if the trespasser decides to carry a gun, or if the trespass is onto publicly owned or prohibited property—an airfield, a military base, a railway, a nuclear power station.

It is a law most robustly enforced, among the ostensibly democratic countries at least, in the United States—and in some particular states, Florida, Louisiana, and Texas, most demonstrably so. It is from states like these that one hears lurid tales of landowners opening fire on uninvited sojourners, even though the law specifically forbids the shooting of a trespasser, unless he is brandishing a weapon and threatening the life of the owner. Warning signs declaring "Trespassers Will Be Shot" are to be seen on all sides in American states like these, and though the signs are permissible as a deterrent, they are not to be regarded as a warning of any impending fusillade.

Where I live in Massachusetts, there is a great deal

of seasonal hunting—for deer, mainly, though black bear on occasion, and with a variety of weapons, including crossbows, black-powder muskets, and rifles, each of which is assigned a specific week in every autumn. Signs at the town limits note that hunters must have, and must carry at all times, written permission from the landowner to pursue their bloodthirsty calling; and during the affected weeks nonhunters are advised to stay indoors and to suit up one's larger pets in reflective orange coats, so that they are not mistaken for deer. In addition, though, the owner must festoon his land's perimeter with orange signs, stapled to a tree every hundred feet or so, with a wordy insistence under the warning word "POSTED," that there be "No Trespassing," followed by a list of specific activities— hunting most obviously—that shall not be pursued.

It seems to most owners that state law tends to favor the hunter: anyone who fails to post his signs the proper distance apart may discover, if the supposed violator has a canny lawyer, that his accusation of trespass is made quite invalid on a technicality. Moreover, if the trespasser injures himself on your property—if he trips and breaks his leg, say—he may have the right to sue the property owner under the principle of *attractive endangerment*, which seems to someone unaware of

America's often byzantine land-related statutes to have turned the law entirely on its head.

In Texas, studded as it is with ranches, especially in the western ranges, regulations for the landowner who is concerned with trespassers are strict, detailed, and much enforced. For instance, under Title 7, Chapter 30, of the state's penal code, which defines criminal trespass as "a person entering or remaining on or in property without effective consent," there are special rules for how such a warning might be presented in unfenced properties, which dominate the state's west-ernmost counties. The caution can be written on

a sign or signs posted on the property or at the
    entrance to the building, reasonably likely to
    come to the attention of intruders, indicating
    that entry is forbidden;
the placement of identifying purple paint marks
    on trees or posts on the property, provided that
    the marks are:
vertical lines of not less than eight inches in
    length and not less than one inch in width;
placed so that the bottom of the mark is not less
    than three feet from the ground or more than
    five feet from the ground; and

placed at locations that are readily visible to any
    person approaching the property and no more
    than:
100 feet apart on forest land; or
1,000 feet apart on land other than forest land

In Massachusetts matters tend to the less formal. It is a commonplace that ancient boundary trees have often grown so much since signs were placed on them by former owners that they have since folded themselves around the old metal plaques that once read NO TRESPASSING but which now have been wizened and their lettering conflated to read NOG or NOSING— which a good lawyer would probably argue renders the boundary invalid, letting any poacher off scot-free.

Warning signs themselves do not keep people off another's land. The American invention that is most traditionally placed to deter intruders from trespassing—and one which has spread worldwide since its invention in the mid-nineteenth century—is barbed wire, *the devil's rope.*

The idea behind an invention that has at least half a dozen claimants to being its originator is timelessly simple: "two wires, twisted together, with a short transverse wire, coiled or bent at its central portion about one of the wire strands of the twist, with its free ends projecting in opposite directions, the other wire strand serving to bind the spur-wire firmly to its place, and in position, with its spur ends perpendicular to the direction of the fence-wire, lateral movement, as well as vibration, being prevented."

The man who, with this elegantly incomprehensible description, lays the principal credible claim to the first patent for it in late 1874* was the son of English immigrants to the United States and named Joseph Glidden. His early demonstration of the usefulness of his creation had one unanticipated consequence: it helped in no small measure to bring about a signal change to the American diet, almost overnight.

The change derives from the simple fact that the first purpose of the wire was to keep animals in, not

---

* A German immigrant named Jacob Haish, who first fenced fields in western Illinois with bushes of the formidably spiky orange osage plant, *Maclura pomifera*, had taken out a patent for barbed metallic wire ten months earlier than Glidden, resulting in a vicious lawsuit that lasted for eight years; Glidden eventually won in the U.S. Supreme Court. *Barbed Wire Magazine*, published in Kansas to this day, writes extensively about fallout from the celebrated case, and it writes fully and sympathetically of the other soi-disant inventors, including the splendidly named Ichabod Washburn, who turned his invention of a wire-drawing process—once he realized he had missed the barbed-wire boat—into the manufacture of sewing-machine needles.

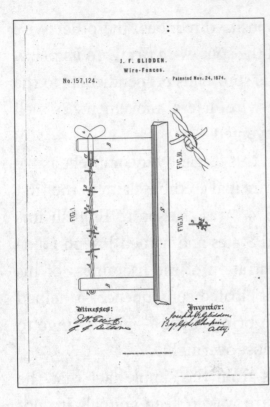

Joseph Glidden's 1874 patent for a sharp-pointed wire confection was designed to keep animals penned in; but as "the devil's rope," barbed wire has since served around the world, mainly to keep strangers out.

to keep people out. And to display how easy this was, Glidden built himself an enormous ranch on the near grassless plains of the west Texas panhandle and housed there the near unimaginable number of 20,000 head of cattle. He was able to corral these animals in such numbers and at such relatively low cost by ringing the entire ranch with his newly made wire—some 120 miles of it, at a cost of some $39,000, far less than a conventional wooden fence, and far less cumbersome.

Having so many cattle pinioned in one place, conveniently close to a railway line that led ultimately to the stockyards in Chicago, played into the great "beef bonanza" that was just then gripping the nation. Beef became all of a sudden both cheap and available, with the result that almost overnight it would replace pork as the preferred national dinnertime dish. Corralling cattle in such numbers became, from the producers' standpoint, economically most advantageous—leading to the invention of that current abomination of the midwestern agricultural scene, the feedlot. Given the known cardiac health disbenefits of today's massive beef consumption—leaving to the side the effects of so unnecessarily large a cattle population on climate change—one might fairly say that Glidden's invention of barbed wire led, in time, to the currently high American incidence of heart attack.

Once Glidden's famous patent, number 157124, had been approved, and with the appeal of his well-publicized panhandle demonstration, so it seemed that every farmer west of the Mississippi was determined to string this newfangled barbed wire along his property lines. The railroads followed suit: not wanting to have livestock, or more especially heavy and locomotive-disrupting bison, wandering dangerously onto their tracks, they also purchased thousands of tons of the wire to spool out alongside their rights-of-way.

After that, for the barbed-wire industry, it was off to the races—with the result that the devil's rope, which over the decades would come in many weights and strengths, with many different designs of barb, leading to today's viciously displeasing sibling *razor wire*, became the world's default barrier to unwanted movement. It kept prisoners in; it kept rabbits (in Australia) out. It helped keep North Koreans from venturing southward, or Pakistanis from attempting sojourns eastward. Coils of it kept Great War soldiers safe in their trenches. And all types of it are on display in museums and at conventions of the various state wire collectors' associations—most notably in California, Kansas, Colorado, and Nebraska—where it is seen as powerfully emblematic of American pioneering and expansion—also being a vivid and potentially painful reminder that to trespass is a most foolhardy endeavor. Especially in America.

But not, it turns out, in Scotland.

For in Scotland there is today essentially no such thing as trespass. One now has an absolute statutory right to wander anywhere in the country at any time of night or day, no matter who the land belongs to, and no matter if the landowner objects to your presence on his acres. Your right of access generally trumps his right to privacy—a revolutionary development that was first undertaken in 2003, and now has the whole world enthralled. Despite furious objection and bluster from some of those who are in possession of many of the Scottish acres, the new arrangement appears in first years of existence to have done very little harm and achieved a great deal of public good.

The right to roam harmlessly across a landscape, to take exercise or simply to recreate the soul, was for centuries an inalienable part of human existence. Common sense and common decency would and should suggest that this right is so basic as to defy any need for explanation. There is, for now, no suggestion anywhere that the general public should have limits imposed on right to breathe the air, nor that one might be forbidden to bathe in the sea. Both belong to all. Land, air, and ocean were once all components of the human birthright—and yet in recent years the public nature of land, uniquely, has been greatly reduced, and common human rights of use of it have been massively attenuated, simply through the introduction of private ownership—and helped by such hostile inventions as barbed wire, warning signs, mantraps, bailiffs, and shotguns.

Pockets of kindly civility remain, though. In Scandinavia most particularly, this ancient right of land wandering—*allemansrätten* is the Swedish word for it, "the everyman's right"—survives and is cherished; and, Nordic peoples being generally much gifted with common sense, it is a right that does not require legislation. The legal principle is simply that of *nulla poena sine lege*—that which is not illegal cannot be punished. The thought of erecting a "No Trespassing" sign is seen in Finland as vulgar, impolite, and quite unnecessary.

There is some regulation, naturally, and some of the rules are intriguing. An outsider may not pick cloudberries in Lapland, for example, because that is something that Lapps do. In Finland, please do not disturb reindeer, steal lichen, or make too much noise when camping close to someone's house. You may however take your horse for a swim in a private lake and you don't have to ask permission. In Norway you may cross cultivated land, but only when it is covered with snow. Anyone who puts up a fence trying to restrict access to a beach there will incur heavy fines. Children under the age of sixteen may fish at will wherever they like, but adults wanting to fish in private waters have to both obtain a license and get permission from the owner. In Sweden you have an absolute right of access to all land, subject only to the restrictive covenant of "do not disturb, do not destroy." Access to Swedish beaches has a handful of restrictions—you do not have an absolute right to walk through what is called the *hemfridzon*, the "immediate vicinity," of a seaside house (though few owners ever get angry if you do), but to give additional benefits to beach strollers, the Swedish authorities try to discourage people from building new houses within a hundred meters of any beach, thus not ever creating a *hemfridzon* that one is supposed to avoid.

The list of explicit roaming rights goes on, bringing some surprises. All of the forest and farmland of Belarus, for instance, is decreed by the national post-Soviet constitution to be publicly owned, and anyone may venture into the country's dark woods and take as much wood, fruit, berries, and medicinal plants as they want and need, without asking. In Estonia you are specifically allowed to gather hazelnuts as you may. In the Czech Republic, on the other hand, you must seek permission if you wish to walk through a garden that grows hops: most other gardens you may pass through freely. Switzerland frowns only on the excessive use of certain parts of certain cantons, and in Bavaria there is a law called the *Schwammerlparagraph*, or the "mushroom clause," which gives all the absolute right to forage for and appropriate wild plants from within the regional forests.

In all of these decidedly non-American places there is an unspoken abhorrence of the very notion of trespass, because—though not to belabor the point—it is so strongly felt that access to land is every bit as much a right as the right to air or water. That was not, however, the right in the United Kingdom until the beginning of the twenty-first century.

Under intense pressure from bodies like the Ramblers' Association, Parliament in London offered legislation—famously the CROW, the Countryside and Rights of Way Act, 2000—that opened up much of the country to more or less the same kind of everyman's right as is recognized in Scandinavia. The government sought a legal means of achieving this—something for which the Nordic countries never felt the need—after accepting that such would compensate for the underlying assumption of most Britons that access to land was *not* a God-given right, but a privilege in the gift of landowners. It will probably take a long while to nudge Britons toward a more Nordic frame of mind in such matters, but London has at least taken a step in that direction.

The Scots, meanwhile, have taken something of a giant leap forward. That Scotland has been permitted to diverge at all from its English neighbors and rulers in matters relating to land stems wholly from the result of a Scots-only referendum that was held in

1997, in which a sizable majority of the country's four million voting residents agreed that Scotland should devolve some its powers from the authority of London, and run much of its own affairs on its own, independently. Accordingly, since 1999 Scotland now has had its own 129-member Parliament sitting in Edinburgh, and although in a further referendum in 2014 its voters opted to remain in the United Kingdom, the country's nationalist leanings have steadily intensified, accompanied by a fiery determination to undertake changes and reforms that remain unpalatable south of the border.

Land reform is one such, and a cascade of legislation formulated in this brand-new assembly, led since 2011 by a nationalist party, has changed matters drastically. The crucial component was an act passed early on, in 2003, that did three things—two of which related to exactly who may buy land in Scotland, and altered forever the curious reality that half of the country's land is owned by a mere three hundred wealthy or land-wealthy families: a curiosity that belongs later in this account. The part of the 2003 act that is most relevant to the notion of trespass was the creation of a legal framework that allows people to have almost unlimited access to all of the land. The fundamental principle now enshrined in Scots law is that all may have access to all land, so long as everyone follows the *guidance*—this cozy and friendly word is used in place of the much harsher word *rules*—of a carefully elaborated Outdoor Access Code. The Scots are proud to note that their code, even though still a code, is much more people-friendly and confers many more rights than is found in the English CROW legislation of three years earlier.

And so—as long as you respect the interests of others, you take good care of the environment, and you take responsibility for your own actions—all Scotland, since 2003, is all yours. Every mountain and moorland, every field and forest, every beach and brae, every loch and lake,* every *sgurr* and scree, is open and available for walking, cycling, swimming, canoeing, horse riding, camping, and climbing. Just no motorized conveyances. Dogs must be kept under firm control—sheep-worrying being a formal term for canine misbehavior for which a dog may be dispatched without benefit of clergy. Otherwise—so long as you do not pass unnecessarily close to a private house nor stray into the

---

* There is, technically, only one lake in Scotland, the Lake of Menteith in Stirlingshire—all other natural bodies of water being *lochs*. There is a scattering of man-made lakes, including a filigree of water in the Lammermuir Hills called Pressmennan Lake, made in 1819, and the Lake of the Hirsel on the estate of a former British prime minister, Lord Hume.

garden—the country is almost as open of access as is Lapland, with the added benefit that such cloudberries as you may find—known in Scotland as *averin*—are yours to keep and eat, since no Lapps with a prior right inhabit the place.

4

The World Made Wild Again

If I breathed the word
That disappeared all people
in the world,
leaving the world
to the world, would you
say it? Would you
sing it out loud?

—SIMON ARMITAGE,
British poet laureate (2019)

Protected by triple fences of electrified razor wire, minefields, and booby traps, wafted across by the beams of searchlights and ranging scans from machine-gun nests, and subject to a litany of treaties and cease-fire agreements, a swathe of four hundred square miles of virgin Korean land has been standing empty and undisturbed since the early autumn of 1953. This dystopian creation, the Demilitarized Zone, the four-kilometer-wide DMZ jagging between North Korea and South Korea roughly along the 38th parallel, is rightly seen as a monument to human insanity, and yet unexpectedly, something quite wonderful and uplifting has occurred within. For by a miracle of zoological and botanical accident, constellations of long forgotten plants and animals have returned to live on this strip in grand feral abundance, lured by the perfect serenity of what is otherwise thought to be one of the most dangerous places in the world.

A living museum of temperate-zone biology has been created inside the 155-mile-long zone. Though it was a barrier specifically designed to separate armies of great ferocity from fighting each other again, it has seen instead a fragile rebirth of natural life, since it is one of the few places on the planet guaranteed to be untroubled by the presence of man. All humankind is forbidden from ever entering the DMZ, and Nature in consequence is gratefully flourishing in the peace.

A sort of peace, at least. Loudspeakers mounted on hilltops and promontories on both sides of the fences blast unceasingly and at unimaginable volume a deluge of propaganda—anthems, speeches, arias, marching songs, Orwellian announcements of victories or outrages—prompting many human observers and sentries to wear earplugs. But the animals inside the wire, if seen through powerful lenses, seem unconcerned: the red-crowned cranes touch down lightly in the thousand-year-old and

Within the 4-kilometer-wide, 155-kilometer-long expanse of the Demilitarized Zone separating North and South Korea live a varied menagerie of animals and birds, untroubled by human presence, which is forbidden. Red-crowned cranes flourish there.

now overgrown rice paddies, and peck and strut contentedly and then take off again, in squadron formation, before settling once more for richer pickings nearby. Other birds are equally untroubled: the white-naped cranes, almost as rare and special, come and go, as do the yellow-throated martens, pheasants and magpies, turtle doves and bulbuls, tits and woodpeckers, all rare in South Korea, but present in what would be (but for the blaring speakers that render them near inaudible) full-throated thousands here. Likewise, the mammals—the Asiatic black bears, the lynx, the Chinese water deer, the goral goats, and the seldom seen Amur leopards, all of them now known to flourish happily in this immense but quite unplanned nature reserve.

The idea of allowing Nature to take over from humankind a tract of landscape and in time to reverse or at least repair the damage done by centuries worth of human activity is not new. As far back as the eighteenth century BCE, the author of *The Epic of Gilgamesh* bemoans the destruction of the Mesopotamian forests, and wonders in the various texts of this long and highly influential poem* how humankind will ever be able to

---

* The Old Testament and the writings of Homer contain stories—Noah's flood a prime example—that first appear in *Gilgamesh*.

answer to God for the wounds inflicted on His world. And today, when we read of Nature retaking the land of the DMZ, of how the irradiated and abandoned territory around Chernobyl has reverted to its previously wild state, we enjoy a certain perverse epicaricacy, knowing that the force that was meant to win—Nature, the jungle—did indeed triumph in the end.

Sometimes it wins so aggressively that it loses, however. Sometime the law of the jungle can prove unpalatably cruel for the humans who chance upon it, preferring their nature to be more pleasing to the eye and mind. Events in a Netherlands experiment initiated in the late 1960s provide an illustration of a collision between well-meaning humans and some of unsettling natural forces that they unleashed.

An earlier chapter discussed the creation of the Dutch polder known as Flevoland, and the manner in which this land, land never before owned by or settled by anyone, first came to be distributed among those who now own and farm and inhabit it. But not all of Flevoland is populated by humans. A section of 12,500 acres, given the name Oostvaardersplassen, close to the capital of Lelystad, was fenced off in 1968 and deliberately wilded—seeded with such animals and plants as Dutch ecologists thought might mimic the northern European marshland the neighboring landscape once had been. Accordingly a number of breeding pairs of large and wild-living herbivores were brought in—red deer, like those in Sutherland; the small and sturdy Polish Konik horses; and Heck cattle, the tough German-bred animals that were the final result of a failed attempt in the 1920s to breed back to that Holy Grail of cattle-engineering industry, the ur-cow, known as the auroch. Every European wilding project wants an auroch, it seems—they have (or so the literature tells us: the last auroch died in a forest in Poland in the early seventeenth century) a magnificent, near regal affect, somewhat similar to the still living European bison or *wisent*, but all efforts to re-create them, in a Jurassic Park–like laboratory event, have failed. So in Flevoland, Heck would have to do.

All went swimmingly at first. The grasses grew, the seabirds flourished, the marshes filled with shorebirds and beasts, and the grazing lands were fully occupied by these newly settled cattle, deer, and horses. There was plenty of contented breeding going on, such that by the winter of 2016, the herbivore population of Oostvaardersplassen had reached a record level of 5230 animals. Passengers on the trains speeding to and from Amsterdam would marvel proudly: a Dutch experiment in wilding, and everywhere there were animals and trees and plants and birds, and all new made on new-made land. The Dutch Serengeti, some called it.

But then came the winter of 2017, and the cold was exceptional. Wilding means: no man-made intervention, no shelter provided, no feed, no medical help. And so, as the winter wore on and the grass died and the land hardened to iron and the drinking water froze too solid for even the most persistent licking, the animals began to die. In their dying, hungry hysteria, they tore the branches off trees, and the trees began to die as well—such that within weeks passengers on the passing trains would see the bloated carcasses of dead deer and cattle lying on the frozen earth, would see horses in states of dreadful emaciation, their rib cages swollen with hunger, limping like Giger's monsters from *Alien*, wandering aimlessly slowly over the blasted landscape of dead and broken trees and scarred earth.

Sharpshooters were brought in to rid the lingering animals of their misery. Wardens came to shift dead animals from beside the railway lines, to limit public distress. By the time the weather began to turn, four thousand of the creatures had died, and there were huge demonstrations and attempts by fearless people— who had to cross high-speed railway tracks and risk the wrath of the law, for this was a wilding attempt backed by the full majesty of the Dutch government, and woe betide anyone who dared to interfere—to feed the surviving creatures, with four hundred bales of hay brought in illegally one dead of night.

A few who had initiated the experiment continued to champion it. Visitors were taken to the seashore and shown the swelling amount of birdlife—sea eagles, for instance, always a hit with civilians, as well as lapwings, avocets, shellducks, and bearded tits. The wretched grazing that had so starved the horses and cows was the result not of overpopulation of herbivores but of the mating success of returning geese, who chomped on the grass in vast numbers, getting to it before the mammals. What took place that winter, the authorities said, may have been distasteful to the taxpaying public—and their visible shocked children—but it had happened according to the laws of nature, and not in the way that human-managed wildlife reserves, so well liked by this same public, like to plan. Wilding can be an unlovely business; oversensitive members of the public should maybe stay away.

In Britain, wilding on a smaller scale has lately developed into such a fascination—or a fad; it is rather too early to tell—that it has become almost chic, widely admired as a reminder that humankind can do something else beside ruin the climate and melt the glaciers and change the weather and generally despoil and pollute the landscape. Wilding is seen as a Good Thing, a means, undertaken by landowners with intelligence and sensitivity, of beating back the tide of ecological disaster that presages, according to a popular book of the same title, the Sixth Extinction, where bees, then insects, then large animals begin to totter, fade, and then vanish altogether, as the world begins to die. Britain is currently going through a period of intense and admirable interest in its surviving wildlife, and books on environmental themes and on individual creatures, sheep to hawks, badgers to capercaillies, are enjoying considerable popularity. In particular, championed most energetically by George Monbiot, a well-respected writer and activist, wilding is seen as a practice likely to help prevent the death of the planet; its practitioners are seen in turn by some as the heroes of the age, our saviors.

The most public figures in recent history of the wilding movement in Britain are Charles Burrell and Isabella Tree, a couple who for fifteen years ran a somewhat conventional dairy and arable farm at Knepp, their estate a few hours drive south of London. It was in 1987 that the then twenty-three-year-old Burrell inherited the 3500-acre estate—which includes a handsome castle, rendering him, unsurprisingly, an

easy target for those class-obsessed Britons who criti-
cize wilding as an activity only for the very wealthy.
Together with his young bride he attempted to make
a go of farming a variety of crops, beef cattle, sheep,
and some 600 dairy cows. But the farm seldom made
money; and when in 1999 a visiting arborist told the
couple that two spectacularly ancient oaks on the property were being slowly killed by the use of farm pesti-
cides, Burrell and Tree made a profound decision, one
freighted with an awful and risky potential: they would
sell off all their mechanical equipment, destroy their
fertilizer stocks, tear down their internal fences, and let
the farm and the creatures and plants on it run *wild*.

They would enforce wilding upon their land. They
would persuade it, via a host of clever and inventive
strategies, to revert to its former and supposedly happier ways. They would *engineer* the wilding of a corner of southern England that they hitherto managed
with so little success. This they would do, among other
ways, by bringing in ancient herbivores once more;
they would employ the advice of the very same Dutch
ecologist, Frans Vera, who had overseen the introduction of Konik horses, Heck cattle, and red deer at
Oostvaardersplassen. The pair would then wait some
years, trying to manage their original dairy and beef
cattle farm within this newly made wilded continuum,

Charlie Burrell and Isabella Tree at their 3,500-acre estate
at Knepp in southern England, which since 2000 has been a
centerpiece of the rewilding movement.

and see if in time Nature could do better at Knepp than
Burrell and his family and their predecessors had ever
managed to do.

But unlike the Dutch cold-turkey, take-no-prisoners
approach, following the pitiless rules of ecology that
distressed so many, some measure of human help was
offered to the animals under the Knepp estate's supervision. Veterinarians visited to monitor the cows'
health. (A herdsman was taken on to search for the cattle who, though more numerous, now had a tendency
to hide themselves deep in the new-growth woods and

among the thickets and bushes that sprang up on the untended fields.) Supplementary food was on hand in case animals needed it, as they had in Holland. Interestingly, though, the animals required little or no extra food and medicine even when stressed—they seem to self-medicate, adapting to their new and unmanaged circumstances. Some say that since no apex predators, like wolves, have ever been stirred into the mix, to make nature flourish as fully as nature intended, the whole affair is something of a stunt—and the fact that glamping sites and yoga retreats are offered only adds to the skepticism some feel.

Not a few full-time farmers disapprove of the practice entirely. Sensible management of land, they insist, is far more responsible than its wholesale abandonment. Or even of its controlled abandonment, which a project like Knepp exemplifies. Americans interested in these much publicized European projects point out that wilding-through-abandonment has been going for decades in the eastern states, as farmers in hard and cold places like Vermont and New Hampshire have traveled south and west in search of more congenial environments, in places like Virginia and Missouri and Oregon where they could find bottomland with better soils and fewer plough-destroying rocks.

The lands they left behind had been well dunged and

fertilized by their livestock,★ so second-growth woodlands flourished, together with groves of sugar maples
that were too old for the wintertime harvesting of their
syrup, but would provide fine hardwood for itinerant
foresters. Aside from stands of ash and hickory and
oak, most of these northern woods were of white pine,
and these grew riotously. The untended streams between these stands were soon stopped up by beaver
dams, the resulting ponds thick with monarch butterflies and perching birds, the woods a wealth of eagles'
nests, armies of turkeys, promiscuous legions of black
bears, coyotes, foxes, owls, porcupines, deer everywhere, and a rich assembly of returned birdlife that may
well have been previously kept away by the din of all
the threshing machinery and stone-wall building and
loud human activity that once marked out a working
farm. What many can see today in rural Massachusetts,
where I live on a long-abandoned eighteenth-century
property, is often classically rewilded land, rich with

★ The serendipity of dung discovery came about in the east of
England when tariffs on Australian wool imports were lifted and
it became uneconomical for East Anglian farmers to raise and
shear their own flocks of sheep. Those running the arable farms
that replaced them found they had bought fields that, unexpectedly, had been amply fertilized by decades of sheep leavings, and
flourished in consequence.

lurking animals and strange sounds, casually untended and with biological consequences unintended and unimagined.

It is not only American farmers who have themselves become a migratory species—all across Europe there are abandoned farms, some of their former owners either having moved to better lands or, more usually, having decided that the prodigious challenges of agriculture are simply not for them and so having fled to the cities. The abandonment of their properties—in Spain, Greece, Sicily, Tuscany—has produced similar results on the land to those in New England, though in many places large sporting estates have sprung up, or Britons eager for sunshine have purchased the properties for a song. These latter are obviously not candidates for wilding—but where a reversion to nature has come about as a consequence of harsh commercial or military realities, the outcome is very different from— and is perhaps more *genuine* than—the well-publicized determination of a group of well-meaning practitioners of what can best be described as rural engineering. Knepp is wilding as theme park: an old Massachusetts property left a century ago by the farming family who built it produces wilding by its own volition, as a consequence of its own abandonment, and not by engineering at all.

Moreover, ask these same farmer-critics: What of the production of food to meet the dietary requirement of the nation—which is, after all, what farms are for? Is it entirely right and proper to permit for prosperous countries in the north to be fed with food grown by faraway people in Africa or South America who may themselves go hungry—one thinks of the unhappy recent creation of a quinoa monoculture in Peru and Bolivia—to meet the demands of the fashionable in the west—while we indulge in the creation of a non-productive wilderness in Sussex? The argument rages still. Should land be stewarded or left to run as it pleases? The fences came down for the Knepp experiment in 1999, the fences had gone up around the Oostvaardersplassen section of the Flevoland polder thirty years earlier. As to whether either is a success—whatever the "success" of such a venture might mean, beyond the possible creation of instructive spectacles for the nature-hungry visitors—is still too early to be certain of. Few are hurrying to copy the example of this unusual corps of wilding pioneers.

5

On Wisdom, Down Under

My mother and my father taught me a lot, how to look after this land. Riding around this country on horse-back we used to go to many places, burning along the way looking after the land. Times have changed, our country has changed a bit, but I still have good knowledge and it is my time to pass on what I have to my children and grandchildren. My mother and father would say if you look after this country, this country will look after you.

—VIOLET LAWSON,
aboriginal landowner, Kakadu National Park,
Northern Territory, Australia (2010)

Joseph Banks, eighteenth-century gentleman-naturalist and a figure of great wealth and social standing, knew enough about land from his own great Lincolnshire estates to recognize through his telescope that something most peculiar was going on, when he first landed in New South Wales at the end of April 1770.

Both he and James Cook, the celebrated captain of the ship that had brought him here, HMS *Endeavour*, remarked on the evidently *managed* nature of the countryside that spread before them. "The land, tho in general well enough clothed," wrote Banks in his journal, "appeared in some places bare . . . with trees not very large and . . . separate from each other without the least under wood." Cook himself remarked later on something even more surprising, especially so to a group of Englishmen who were now twelve thousand miles and two full years' sailing from home. After an excursion into the countryside, he wrote, "we found it diversified with woods and lawns and marshes; the woods free from underwood of every kind and the trees are at such a distance from one another that the whole country or at least a great part of it might be cultivated without being obliged to cut down a single tree."

And to add to the oddity of the situation an expedition member named Sydney Parkinson, whom Banks had employed as his draughtsman and who created hundreds of sensationally beautiful—and scientifically accurate— drawings of plants that his master had discovered,* remarked that from a distance this new land "looked very pleasant and fertile; and the trees, quite free from underwood, appeared like plantations in a gentleman's park."

Clearly these were improvements, and they had been created by human agency. It did not take long to find out who. *Endeavour* was at the time closing fast on an inlet in the steep cliffs at the edge of the territory known then as New Holland, named for the Dutchmen who had first seen it but never explored it. Crewmen up in the crow's nest suddenly spotted people—two men with spears and sticks keeping watch from the rocks at the bay's entrance, a group of four others fishing unconcernedly. Cook turned his vessel, the sails caught

---

* Sydney Parkinson's drawings and watercolors were eventually engraved onto copper plates and printed into a thirty-four-volume *Florilegium* which was published in London between 1980 and 1990. Parkinson never got to appreciate the massive contribution he made to botanical illustration, as he died of dysentery on *Endeavour*'s way home, between Java and Cape Town. His work was conducted under trying conditions in a cramped shipboard cabin; he complained that while in Tahiti, insects ate most of his paint.

the breeze, and in swept his little ship. A group of sailors, the captain leading them, then piled into a pin-
nace and headed for the shore. There were half a dozen
small bark tents on the foreshore, and beside them a
group of dark-skinned women tending a fire. The four
fishermen Cook's crew had seen earlier returned and
gave the fish to the women.

Cook sensed this might be a good time for a visit—
but was instantly rebuffed, with men armed with
spears making an angry run at the visitors. Cook tried
to calm then, threw gifts down onto the beach, only
leading to more cries of evident hostility. In an attempt
to induce—the phrase reeks of imperial hauteur—a
change of attitude, Cook had one of his men shoot
(lightly!) at the most aggressive of the group—a man
who instantly ran off to his tent, dropping a shield as he
ran. That shield was brought back to England as a spoil
of war, and is now in a museum in Cambridge; descendants of the victim, now known to have been a member
of the Gweagal family of Australian aborigines, repeatedly visit Cambridge these days in an attempt to have
what they consider a stolen object, returned. Thus far,
nothing doing.

In the end Cook and his crew, accepting that their
first contact had been less than a consummate success,
gave up the effort, retreated to the ship, and set sail

The first encounter between Captain James Cook and Australian aboriginals in 1770 was not a happy one, resulting in a skirmish from which this wooden shield was taken, as a spoil of war. It is now displayed in a Cambridge museum, and its owners want it back.

northward along the coast. Inside the Great Barrier Reef they ran into difficulties when their ship almost tore her hull apart on a coral spur. They had to careen her and make repairs—eventually waiting for some seven weeks at the mouth of a stream they named for their craft, the Endeavour River. And the settlement that stands there today is, unsurprisingly, Cooktown.

It was near here that two further matters transpired. James Cook formally claimed the newly found land for England, and named it—the vast entirety of the east coast of what had first been New Holland—New

South Wales. And he and his crew made passing good friends with the local aboriginal people—a very different group from the Gweagal down near today's Sydney, and named the Guugu Yimithir, who proved helpful and supportive during the careening sojourn. These people, Cook and Banks later observed, were impressively diligent in their care for the countryside and took great care with their plantings and their management of the local forests. They evidently had a sophisticated understanding of the natural world and its processes. All of the Englishmen were impressed—though it remains a cruel irony of the expedition that the flattering remarks about the Guugu Yimithir were all excised by the Admiralty from the official report of the venture.

Doubtlessly, Cook, Banks, and the officers at that point would have been reminded at this of the official advice given to the expedition before it had set out from England two years previously. Their voyage was sponsored by the Royal Society* and the body's president,

---

* The official stated purpose of Cook's expedition was purely astronomical, to observe the transit of Venus across the face of the sun—an observation that, if sufficiently accurate, would allow by triangulation a fairly precise determination of the distance between the Earth and her parent star. The detour to Australia was officially only incidental; unofficially, it was the key component of the trip.

the Scottish astronomer James Douglas, fourteenth Earl of Morton, had slipped a vade mecum into Cook's hand as the ship prepared to depart. Should there be inhabitants of the lands you find, he said,

> they are the natural and in the strictest sense of the word, the legal possessors of the several Regions they inhabit. No European nation has a right to occupy any part of their country, or settle among them without their voluntary consent. Conquest over such people can give no just title; because they could never be the aggressors.

That James Cook so blatantly disregarded these decidedly nonimperial instructions—first, by seizing the country for the Crown and second, by giving such scant regard to the wishes of the locals—casts an evident blemish on the reputation of the man. In terms of the relationship between white settlers and millions of aboriginals in the colonial and postcolonial years to come, Cook's actions helped to establish a legacy of cruelty, carelessness, and arrogance that has marked the relations between these two groups to this day. The aboriginal civilization, one of the world's most ancient, was disregarded for generations, its people reduced in circumstance and stripped of all their dignity.

Except that every so often, the skills, knowledge, and reverence for the land that characterizes so many aboriginal people in Australia, and which so keenly impressed Joseph Banks and his colleagues at the end of the eighteenth century, reassert themselves, reminding the world of the unusual qualities of an otherwise somewhat overlooked people. These unsung and unremembered qualities came to the fore most recently in early 2020, when much of Australia was consumed by bush fires of unimaginable ferocity and destructiveness. The fires exerted a terrible toll on the psyche of the white Australian people—and it caused them some pause when aboriginal leaders reminded them, and the world beyond, that fire itself was perhaps not so malign a phenomenon. Rather, it was something long employed as a tool of ancient agriculture in the continent's history, and that perhaps the white settlers could learn a little from those whom legions of white Australians had long dismissed as mere "blackfellas," when in reality they patently still knew so little about the land on which they had settled. In fact, as Cook and Banks had supposed as long before as 1790, the aboriginals had a profound knowledge of their countryside, and they well knew how to treat with it for the good of all who lived on it.

Fire, which today seems to be the ruin of modern Australia, has since ancient times been employed to encourage society's salvation. More than a few times during the devastating midsummer weeks of 2020, with cascading conflagrations propelling southeastern and southern Australia into the world's headlines, aboriginals would shake their heads in sad bewilderment and mutter that, essentially, *we told you so.* For they had long practiced the controlled burning of underbrush— the *underwood* mentioned earlier, whose absence Cook, Banks, and Sydney Parkinson had all noticed two centuries before as unusual—as a means of ensuring that larger fires, whether caused by dry lightning or by carelessness, had less of a chance of spreading, of getting out of hand. What puzzled the native people was that under white supervision so much woodland underbrush had been allowed to grow unchecked— and that in a furiously hot country, where sooner or later fire would be bound to break out and would then, with so much fuel carelessly left around, prove mightily difficult to control.

Violet Lawson, an aboriginal activist who lives in the small town of Jabiru in the far north of the country, is one of many who teach the practice of what is called *stick-burning,* to create what are called *cold fires* in the immense eucalyptus woodlands where she lives. Come springtime and she'll be out in the woodlands lighting hundreds of fires each week, in small patches, the timing of the burns dictated by the air temperature and the wind direction. She'll carry a lighted piece of shaggy bark with her—much as a Scottish farmer might carry his fire-lighter to burn off the wild heather up in the Highlands—and every so often she will brush it down to smear the flames onto the grass and so set ablaze small tiles of suitably flammable patches of grass and brush, watching them and tamping them down if they ever seem likely to spread beyond her line of sight. The idea is that the fire will be cool enough to preserve any sensitive plants, yet hot enough to burn off the dead and unnecessary foliage. According to an aboriginal forest ranger nearby, the set fires should tickle, not rage.

In a way the technique is not too different from the lawn mowing of suburbia—it both keeps down excess untidiness, and it encourages regrowth, in this case by allowing a thin cover of nutrients to blanket the soil on which new blooms of grass seeds will inevitably fall. The technique in this remote and thinly settled part of the country has long been to set small fires cleverly in order to prevent big fires. By contrast the response among the supposedly more sophisticated white inhabitants in the urbanized corners of Australia is simply—

and expensively and life threateningly—to wait for big fires to break out and then to marshal small armies of men and machines to join battle with them.

The Australian government, or more properly some of the more forward-thinking Australian states, have evidently been listening, and in more than a few communities today the use of traditional aboriginal fire techniques are being permitted and even encouraged. This is a profound reversal from earlier colonial attitudes, which actually forbade native burning—early

Indigenous Australians have long practiced the technique of setting regular and well-controlled "cool fires" to clear underbrush and minimize the likelihood of major conflagrations.

settlers had seen the distant landscapes of their new homeland pockmarked with rising trails of smoke from aboriginal cold fires that had been set deep in the woodlands. They had complained, and had done so with sufficient vehemence to have what they in their ignorance considered a dangerous practice totally banned. Similar Victorian-era bans were also enacted in North and South America, Africa, and in various parts of European-controlled Asia: natives armed with fire were seen as threatening, and skittish Europeans were scared to death by them, not knowing what the fires were for, assuming sinister purpose, and worse.

It is a curious reality that many aboriginals who visit Australia's southern city suburbs today say that it is they who are scared to death, and by the very opposite—by the existence of the thick and untended woodlands and the unburned stretches of underbrush and high and dry grassland. "I was terrified," remarked a young indigenous woman who had settled briefly in the dry-country town of Nowra, two hours south of Sydney. "I couldn't sleep. I told everyone that we need to go home. This place is going to go up, and it is going to be a catastrophe."

And indeed, in Nowra, her apprehensions were amply justified. A local newspaper from January 2020 reported on just the kind of catastrophe that the woman had feared, a cruel classic of Australia's most recent bad fire season. The town was consumed by flame. There were fires raging to the north of the small town. huge fires to the south, flame heights of at least twenty meters rising up on one side of the Shoalhaven River and forty meters on the other, and dozens of smaller fires in areas that hadn't burned in decades. An evil-looking yellow, greasy-seeming pyro-cumulous cloud rumbled with its self-generated thunder outside. "The southerly is going to hit around nine, 9.30 and unfortunately you're the ones that will have to deal with it," the local fire chief was quoted in the paper as saying, trying to explain the situation to anxious townspeople. "What it's going to do I have no fucking idea."

He really did have no idea. The southerly winds hit two hours earlier than expected, roaring through the forest with gusts of more than one hundred kilometers per hour, whipping up dirt and dust and ash and filling everything with smoke. Strike teams rushed here and there, trying to douse fires that sprang up almost at random. At Nowra a firefighter explained what he thought of as strategy: we can't stop the fire, so we're just trying to direct it as best we can. Strike teams were sent across the state on Saturday, moved around like chess pieces against an opponent that ignored all the rules. "It's turned to shit everywhere."

But it needn't have done so.

An academic paper devoted to the aboriginal practice of controlled burning, and published in 2010, with Jabiru's Violet Lawson one of the nine authors, made an important point about the attitude of white Australians to the fires that so frequently afflict them. Indigenous peoples have had centuries of experience with fires, she wrote, and they generally should be listened to, their wisdom and knowledge respected and absorbed by those who have long supposed them to be mere primitives.

Indigenous people generally are all too often generously armed with foresight and wisdom about the management of their lands. One example far away from Australia has been quoted over and again: the Andamanese Islanders, who live remotely and have had little contact with "civilization" in the seas off southeastern India, astonished all by surviving en masse the devastating 2004 tsunami, which killed a quarter of a million people elsewhere. The islanders had weathered the giant waves in large part because their ancient songs, handed down through the generations, had instructed them that *when the seawaters suddenly receded, they should run and take to the hills*, which they did—and all escaped drowning.

The fact that these islanders did not have an advanced written language rendered them, in the eyes of their supposedly wiser white neighbors, unlettered and so merely primitive. But they possessed, unknown to most outsiders, a rich and powerful oral tradition, and that very tradition provided them with sufficient a supply of wisdom to outrun the deadly waves that December day. Their survival briefly reminded the outside world that it was those who lacked such prescience who lost their lives. They turned out simply to be too clever to understand the inherent dangers of a wild and capricious Nature.

As with the Andamanese, so with Australian aboriginals. The academic paper spoke admiringly of these long-unrecognized skills of theirs, and of the pressing modern need both to respect and to pass on to others what they knew, and we did not.

Aboriginal people have occupied northern Australia for at least 40,000 years, and over this period have developed a rich culture of law, ceremony, oral history and detailed ecological knowledge. Despite nearly two centuries of European colonization, large areas of northern Australia remain in Aboriginal ownership or have recently been returned to indigenous management and control. A high priority for Aboriginal people is to record and revitalize their indigenous knowledge and practices to meet stewardship obligations and to ensure they are available for younger generations of Aboriginal land and sea managers. In recent years there has also been increasing recognition by non-indigenous peoples of the value of applying such traditional ecological knowledge and practice to contemporary land management.

In late November 2018, a group of aboriginal fire crew members were found conducting a traditional coldfire burn close to a country town even farther south of Sydney. A few days beforehand this community had suffered horribly from a disastrous blaze which had torn through three thousand acres and destroyed more than a hundred houses. But where the indigenous teams were working, five miles away, teams were patiently using their ancient techniques on a stretch of unburned gum-tree woodlands, setting small fires here and there to clear away the underbrush, smearing the fire onto the underbrush, stamping out the flames if they ever became too intense, and doing so with almost Zen-like calm. No shouting, no panic, no heavy machinery. Just a shaggy-bark firelighter and a sense of casual purpose, each member of the team seemingly working at one with the woodlands, at one with Nature.

One young aboriginal firefighter, George Aldridge, spoke almost lovingly of the surrounding landscape as he applied his torch to patches of grass here and there, watching the flames grow and then die away, after which he would sift through the warm black soil left behind and imagine how soon fresh grasses would grow back. "I love land management and looking after my land," he said, kneeling among the wisps of fresh and fragrant blue smoke. "The land is my mother, our mother, and she looks after us, and so it is only right that we look after her."

Except that, all too often, we do anything but.

6

Parks, Recreation, and Plutonium

I live not in myself, but I become
Portion of that around me; and to me
High mountains are a feeling, but the hum
Of human cities torture.

—LORD BYRON,
Childe Harold's Pilgrimage,
Canto III, St. 72 (1816)

C ities are where land comes to die. Maybe this was not so at first, back when the gatherings of large numbers of people managed to enjoy a coequal existence with the countryside around them—in early centers of civilization like Ur and Babylon, Haarlem and

Taos, and Angkor and Teotihuacan. But come the late eighteenth century, and the Industrial Revolution, and the massive global transmigration of populations from *rus* to *urbs*, so the land itself began to vanish, the soft beauty of nature devoured by the hard manifestations of man. Some few mourned. "They paved paradise," went their song. "And put up a parking lot."

In a legal sense the land continued to exist, of course, if now largely underneath and invisible. And in terms of its value or its cost, and even to some its worth, land became in cities a thing of very much greater price. But what had long made land *land* was now steadily changed, with city land rendered into an entity quite different in style and substance. Today, where great cities stand—Tokyo, Mexico, Shanghai, London, New York, Cairo, Los Angeles, Chongqing, Seoul—pastures and forests have been replaced by asphalt and concrete, green has given way to gray, embanked streams have become tiled sewers, burrowing animals have become subway trains, valleys between mountains are now car-clogged canyons separating skyscrapers.

In the suburbs beyond the urban limits, the degradation of the land has been more insidious, its demoted status often cunningly disguised. Such land as appears to exist is mostly artifice, a simulacrum of countryside, the greenest of its expanses available at great expense to the golfer or more ironically to the members of what for the past two centuries have been called *country clubs*. These last are institutions placed well beyond the real country they seek to resemble and offer a reminder—for a considerable annual fee—of the rural dreamland that some old-timers recall went before.

The public spaces of cities, where they exist, have long been an acknowledgment of this loss, presenting a substitute. The agora of ancient Greece recognized the need for a space where all could come, to hear the addresses of their leaders, to meet with one another, to indulge in politics, or to offer merchandise. The agora was, in effect, the public commons; and in the centers of some cities still common land remains preserved, religiously: in Newcastle-upon-Tyne in northern England, for example, there is the Town Moor, more than a thousand acres of preserved agricultural land far larger than the great artificial parks of London and New York, and where some may graze sheep and cattle, course rabbits, and, at one time, take off in small aircraft.

Stanley Park in Vancouver is almost exactly the same size, and it enjoys with the Town Moor the benefit of being essentially unaltered by human agency: it was once a military reservation, bought by the city in 1886 and left as the small garrison of artillerymen kept it, a wild forest of cedars, spruce, and hemlock standing between the city and the sea. The city has seldom employed landscapers to prettify the park: it can almost lay claim to being an island of the rural original, now circumscribed by the new and the man-made.

Municipally owned and landscaped parks are by far the more frequently encountered: the two great Napoleonic hunting grounds, *les bois* of Vincennes and Boulogne on the eastern and western edges of Paris, respectively, are still suggestive of real countryside; the great parks in Los Angeles and San Diego, as well as in Richmond in London, have been carefully maintained and sculpted to help calm the nerves of the frantic city dwellers nearby, though the notion that they are in any sense common land, fully available to all, is misleading. Richmond, Bushy, and Greenwich Parks in London all have considerable populations of deer, both red and fallow, but they are far from being as wild as they look. The animals' forefathers were introduced there for sport by King Henry VIII, and the lands on which they now flock are properly examples of what may once have been common land (Bushy Park enjoyed that status six hundred years ago) but are technically part of the nation's royal estates, and such animals as live there belong not to the common folk, but to Britain's reigning king or queen.

And then there are the *maidans* of eastern cities,

large expanses of meadowland that act, literally and metaphorically, as urban lungs, offering well-ventilated respite from urban congestion and in older times, the risk of tuberculosis. Some of these—the Kiev Maidan in Ukraine and Tehrir Square in Cairo—have lately become notorious as centers of political activism, where police and protesters have been joined, all too often with lethal result. Others, like the Maidan in Kolkata, which stretches unencroached and cherished between the Chowringhee district and the Hooghly River, have a more curiously postcolonial standing, seen by today's Bengalis almost as a compensatory apology for the foreign imperial authority that went before. Its vast expanse was constructed by the British when Calcutta was their Indian capital, and it once doubled as a parade ground and, more sinisterly, as a free-fire zone that protected the Indian army's Fort William from the threats of bandits—soldiers would in theory fire down from the high fort walls at any approaching rebels. Now, though, and with the imperialists gone, the thousand acres have been returned almost to wildness, with grazing goats and cows, with vultures and kites and kite fliers, with cricketers and soccer players and strolling lovers. The land still all belongs to the army, though naturally, since 1947, it is the army of independent India. The land could have been subdi-

Originally established by the British military to protect their Fort William barracks, Kolkata's Maidan offers blessed rural relief for the millions living in one of the country's most overcrowded cities.

vided, with houses provided for some of Calcutta's teeming millions. But it remains untouched, almost penitent, a poignant reminder of what for Indians were unhappier times. Except that, curiously for visitors, the Victoria Memorial, an enormous and triumphal confection of marble made on the orders of Lord Curzon, still stands untouched on the Maidan's southern side, one of the few memorials to colonial times that survives in modern India. Its sheer existence, unscarred, somewhat mutes the pleas for forgetting and forgiving that the Maidan otherwise suggests.

Forgiveness would also seem a proper plea for the

planners of all too many other of the world's cities, however. For while there are legions of truly grand and beautiful cities—even if so many today are greatly ruined by the press of tourism, with Venice and Dubrovnik at the head of the league—they are very much a minority. The greatest number of those cities where the world's millions are now compelled to live, most of which are new and have been built in response to population growth and upon land quite recently pristine, can fairly be described as awful.

These are today's cities of "dreadful night," as James Thomson's poem once described mid-nineteenth-century London, and which Gustave Doré so memorably drew. For every properly planned and hopefully built would-be utopian city—Welwyn, and Port Sunlight, in England; Chandigarh in India; Islamabad in Pakistan; even New Harmony in Indiana—there are a score of ugly, overcrowded, and ill-conceived aggregations of humanity, with few redeeming features and all too few reminders of the natural landscape they replaced. There are over one hundred cities in today's China that have populations above one million souls: most have been occupied in the last half century, few in all candor can be described as pleasing places in which to live. Similarly, India: of the world's fifty most polluted cities, fourteen are in China, and one each in Mongolia and Bangladesh, but the remaining thirty-four are in Pakistan and India. Not that the west gets away scot-free. Many will recall the truly shocking conditions of the housing for America's urban poor—the unutterably terrible Pruitt-Igoe housing complex in St. Louis, for example, now mercifully torn down, or the contrasting city miles of Jefferson Avenue in Detroit—with until lately utter urban wreckage at its western end and then, once over a canal bridge, the mansions of Grosse Pointe where one could live only if one's "degree of swarthiness" was found acceptable.

The generally no longer legal practices of redlining—where credit is limited to certain neighborhoods—and of restrictive covenants, legal fictions that discourage people of certain races or habits from living in certain places, are features of the recent urban world that are well beyond the scope of this account. As are the principles of eminent domain—*compulsory purchase*, in the phrase beyond America—which can quite literally allow the bulldozing of inconveniently placed houses (homes to some, of course) in the questionable name of progress, as for the making of a highway overpass or the building of a commercially important shopping mall. Such practices have savaged many parts of many cities, too. But all these legal maneuvers, if outside the scope of these pages, still serve to remind us that all

A sixty-acre parcel of land in St. Louis was set aside in 1951 for the creation of one of America's largest public housing projects, the Pruitt-Igoe Apartments, which were considered such a crime-ridden social failure that all were demolished after little more than twenty years.

are constructs of modern, civilized, western man: no such limitations were ever a feature of Native American life, nor were they or are they to be found among the Yanomami or the Fuegian or the Bantu, the Inuit or the Ainu. What we do to ourselves as humans we tend to do most egregiously in our cities: though we behave poorly out in the wilderness—think the wholesale confiscation of Indian lands, the annexation of vast acreages for purposes of which in hindsight few of us can be proud—it can be argued that the sheer presence of landscape inhibits our baser instincts to pollute, to make ugly, to ruin. At one time or another, one imagines, even the coldest-eyed lawyer and most stone-hearted developer might stop to consider the view of a range of hills or a glade of trees or a flight of passing geese and wonder: Do I truly need to make this *here*, and spoil so much that was once so fine?

It is easier to do this in, or near to, our cities. They are already ruined, so why not add another layer and make still more money while the going is good? One particularly egregious example of such behavior is to be found in the so-called Mile High City, Denver, in the pretty eastern foothills of the Rocky Mountains.

Nearly every American city of any size is encircled by a highway originally designed to help speed traffic around its central metropolitan area. London has the M25, Paris its Périphérique. But most commonly such roads are an American phenomenon, and some have become famous, part of the national lexicon—Washington's Capital Beltway being a fine example, political sophistication within it, costly suburban sprawl beyond. Similar circumferential highways are to be found across the country, from Atlanta to Fort Worth, Minneapolis to Charlotte, St. Louis to Louisville.

But there is not such a roadway, as any map will

show, around Denver. Although there is a toll road, Route 470, that runs around the city's northern, eastern, and southern sides, and which as it circles keeps more or less a dozen miles out from the office towers

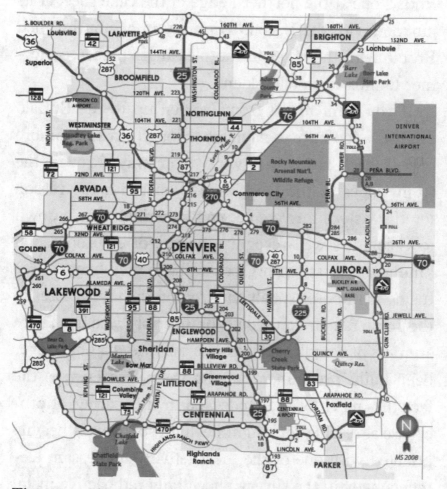

The expressway surrounding the city of Denver, Colorado, sports a large gap in its northeastern quadrant, largely because of fears of plutonium contamination in the soils.

of downtown, it inexplicably vanishes on the Rocky Mountain side of the city. From the air, and on maps, it looks rather like a cake with a triangular slice taken out, and for years those living in Denver have been frustrated by two huge yellow signs proclaiming "Freeway Ends," one at the northern edge of the cake slice, at its junction with the Buffalo Highway in a suburb called Broomfield, the other in the pleasingly preserved former mining town of Golden.

If you want to get from Golden up to the Buffalo Highway you have to navigate through a maze of more or less back roads that wind their way through the front slopes of the Rockies. If you think the reason for the lack of any connecting superhighway is down to Colorado's well-recognized sense of environmental kindheartedness, with green-minded tree huggers refusing to impose asphalt and iron on the virgin acres of the front range, you would be horribly wrong. There are rafts of complicated reasons that the road has never been built: preeminent among them, however, is the fact that much of the landscape in this corner of otherwise agreeable-looking mile-high suburbia is heavily polluted with one of the most lethally unpleasant elements known: the silvery and highly radioactive metal first discovered in 1940, plutonium.

For forty years, from 1952 until 1992, this was

the secret and tightly guarded site of the Rocky Flats Plant, a sprawling factory which manufactured components that would be assembled into atomic weapons. Specifically, the factory and its five thousand workers were charged with making plutonium pits, each one the bowling-ball-sized heart of a nuclear weapon. Plutonium pits are to be found in all three kinds of atomic weapons: the standard fission weapons that have been used since the first tests and their subsequent employment in Japan in 1945; the boosted fission weapons that were tested during the five years after that, through 1950; and the two-stage thermonuclear weapons of terrifying power that were first tested in the Marshall Islands in 1952. In the first two types of weapons, the plutonium pits *were* the exploding parts of the bombs; in the thermonuclear weapons, the pits were the triggers, whose detonation would produce massive explosions in themselves, but explosions that would in a split second activate the fusion component of the bomb and create the final almighty detonation. Once the technology for these types of weapons had been refined, it became clear that huge numbers of pits needed to be made—and a strategic decision as to their manufacture was made.

This decision held that the various parts of America's enormous arsenal of nuclear weapons should be made in factories that for security reasons would be widely dispersed across the United States. The plutonium itself would be made in Washington State, at the Hanford reservation, or later in Tennessee, at the Oak Ridge plant. The weapons would be assembled at the Pantex Plant in western Texas. The pits, made of metal produced in Tennessee, would be machined here at the Rocky Flats Plant in Colorado, to critical tolerances displaying near perfect precision, and to designs that had been created in the two principal government nuclear laboratories, in Livermore, California, and Los Alamos, New Mexico. Once the pits were machined, they would be taken, under conditions of heavy security, to the Pantex Plant near Amarillo—and the finished weapons would in due course be sent, in convoys of booby-trapped and highly secure tractor-trailers, protected by legions of soldiers and airmen and overflown by helicopter gunships, to the various bases—naval or air force—where they would be loaded onto their various types of delivery systems, to be then armed and ready for action.

The Rocky Flats Plant was established to deal with the dirtiest part of atom bomb production—the molding and machining of the truly horrible and impossibly dangerous radioactive metal that lay waiting to fission, the black heart of the weapon. The site on which this work would be done, 6400 acres of gently undulating foothills countryside some fifteen miles northwest of Denver—almost exactly where a beltway should eventually be constructed, but never has been—was acquired by the government in 1951. The Dow Chemical Company was hired to operate the first buildings where the plutonium would be smelted and forged and machined. Within six years there were twenty-seven buildings on the flats; by the 1960s, there were twice as many. All the tasks required to fulfill the needs of the armed forces would be carried out under conditions of the tightest military security and with supposedly hermetically airtight environmental controls. No terrorist should ever be able to lay his hands on an American atomic bomb component. Not even one microscopic morsel of plutonium should ever be allowed to escape from the buildings that were now sprouting up all over the site, and make it into the atmosphere, onto the ground, or into the water. Much would be doomed if ever such came to pass.

And yet of course, it did. It did first of all under Dow Chemical's twenty-six years of supervision—there were at least two devastating fires involving weapons-grade plutonium, most particularly the isotope 239Pu. This form of the metal has to be one of the most dangerous substances known. It is highly radioactive. Its radioactivity has a half-life of 24,110 years. If you fashion it into a ball four inches in diameter and weighing a little under eleven pounds, it has the potential to go critical and explode. With sufficient sophisticated engineering—although the kind of sophistication only available to a sovereign nation—a critical mass can be reduced to the size of a tennis ball. Pieces of it are formidably difficult to machine: it catches fire with consummate ease, it burns ferociously hot. It did so in the two Dow Chemical fires at Rocky Flats, burning through both the gloves used to handle it and the Plexiglas shields of the chamber in which the operator was trying to cut and shape and polish it. On burning, the metal releases—and out into the atmosphere, if not caught by scrubbers—millions of particles that if breathed in will cause (not *can* cause; *will* cause) human cancers so quickly that it has been calculated that a pound of the material will easily kill two million people.★ Plumes of the burned plutonium escaped from chimneys after those fires in 1957 and 1969—the latter taking two years to clean up—and were spread by the prevailing westerly downslope winds onto the

---

★ Ralph Nader, the consumer activist with a vast following, suggested it would kill eight *billion*, but his arithmetic was later challenged and found sorely wanting.

Plutonium triggers for America's growing arsenal of atomic weapons were made at the Rocky Flats plant outside Denver until a pollution scandal closed the site in 1992. Years of cleanup followed.

fields and farms and tract-house lawns of western Denver. In 1972, and to help contain any further spread of contaminants, the federal government bought a further five thousand acres as a buffer zone around the plant, then five thousand more acres a little later. Problems within the plant began to spiral out of control—leaks were found in barrels of radioactive waste, elevated tritium levels were found in local water supplies, windblown plutonium was found in local soil samples, health officials in Denver began to complain, housing

developments in the western suburbs were slowed—
and soon after that the protests began. Hundreds and
then thousands of anxious Colorado citizens were at
the gates, some breaking in and getting themselves arrested. Some local landowners sued the government
for contamination of their properties. All adding to the
publicity of a plant that was fast going rogue.

Dow Chemical was relieved of its duties in 1976 and
Rockwell International—a once vast conglomerate involved in a spectrum of manufacturing, ranging from
space shuttles to television sets, truck axles to yachts,
GPS devices to coffeemakers—was given the new contract. From 1975 until 1990, the company could add to
its portfolio of products the plutonium triggers for hy-
drogen bombs—the making of which required a set of
tasks and skills in which the company turned out to be
utterly lacking. In fact, Rockwell proved so incompe-
tent, corrupt, and irresponsible a manager that in 1989
something quite unprecedented took place, all the more
remarkable considering that the Rocky Flats plant was,
essentially, a federal agency, part of the atom-bomb-
responsible Department of Energy for which Rockwell
had been given operating responsibility.

What took place on the morning of June 6, 1989,
will long stand tall in the annals of the American bureaucracy. A team of FBI agents, cleared through the

plant's formidable security barriers, arrived for an ostensibly routine talk with plant bosses about putting in place protocols to deal with any potential terrorist threats. Perfectly ordinary. No drama. Just normal.

But suddenly, after the managers were all comfortably seated, the FBI leader revealed just why they were actually there. He formally announced that This Is a Raid. It was the beginning of a criminal investigation that had been given the official name of Operation Desert Glow. He came armed with a search warrant, which he promptly took out of his briefcase and laid it on the table in front of the suddenly shaken, instantly white-faced managers. Federal agents, he said, were at that very moment piling out of buses to break down every door to every one of the 800 buildings and raid every filing cabinet and open all cellars and use crowbars to crack open storage facilities, and all to find out exactly what was going on inside the Rocky Flats factory—because it was suspected that at the very least the country's pollution laws were being systematically and seriously violated, and that legions of criminal conspirators were busily covering up breaches of the law that had been occurring for many years previously.

One federal agency was, in other words, investigating a brother agency, and on charges of criminal incompetence and collusion. Such a thing had never happened before. But the FBI, it turned out, had ample reason for doing so—its agents knew well what they were doing: for over the previous twelve months they had been conducting surreptitious surveillance, flying planes over the plant to gather radiation evidence, asking whistleblowers to step forward, to bring photographs, recordings, and leaked files. A grand jury was convened. Charges were brought, deals were done, secrecy was generally preserved (except for a superb exercise in investigative journalism performed by a then stripling magazine called *Westworld*, which blew the lid off what could have been a major judicial scandal as well), and Rockwell was forced to pay some millions in fines. Senior managers escaped criminal conviction and jail sentences, and the whole affair left a lingering impression of justice denied in the vaguely expressed interest—this being the atomic bomb industry—of national security.

Then, in the 1990s, Rockwell was sacked and the plant was scheduled to be closed down. The U.S. navy announced it no longer had a need for plutonium triggers for its latest generation of submarine-launched missiles, and so the business of producing pits at Rocky Flats vanished, almost overnight. Four thousand workers were laid off, a further four thousand were transferred to what became the principal business of the facility—cleanup. Plutonium was shipped off to other plants, waste material was set in concrete and shipped elsewhere, and the multiyear billion-dollar program of remediation and restoration so familiar to other nuclear sites—Hanford in Washington, Three Mile Island, Chernobyl, Fukushima—got ponderously under way. By October 2005 it was declared that, to the best of the federal government's knowledge, all was clean and uncontaminated once again, and life would be allowed to resume. Except—no digging below three feet deep, since only the upper thirty-six inches of soil in Rocky Flats had been fully cleaned or replaced.

There have been lawsuits aplenty in the years since, and monies have been awarded—$375 million in one case—to homeowners claiming that plutonium pollution has spoiled their health or lowered the value of their homes. Part of the site has been made into a wildlife refuge, the opening of which was much delayed and which many Denver parents will not allow their children to visit. And at the time of this writing, the town of Bloomfield says it is not permitting an extension of the Denver Beltway, the long-interrupted E-470, to pass through its section of the former Rocky Flats site. Residual radiation is too high, the town insists.

The stewardship of land is a complicated business— and enclosure or clearance, fencing or wilding, all have their adherents and their critics. But none of these phenomena have, for good or ill, a permanent or even a very long-standing effect on the land itself. Fences are built and then come down, ownership changes, rules relax or tighten, populations ebb and flow, wild animals and plants populate and decimate, then populate again, wildfires rage and seeds sprout once the flames have died down—landscape is all a part of an ever changing world, and it forgives or forgets almost all of the assaults that mankind willfully or neglectfully imposes upon it.

Except, as it happens, in the matter of radioactivity. The soils of the front ranges in this beautiful part of the state of Colorado are now speckled with traces of plutonium 239. This diabolical man-made substance will remain in place for thousands upon thousands of years—a near permanent stain on the planet, a reminder of how cruelly mankind deals with his lands and his landscape.

Colorado happens to be one of the states in the union with the fewest Native American reservations, having no significant number of resident tribal peoples penned up on federal lands. Once there were Cheyenne and Arapaho, Apache and Shoshone. But no longer. Perhaps it is as well. Considering how the invading white man has dealt with the land that the native people have so long revered, they would all in concert bow their heads, and weep.

PART IV

Battlegrounds

1

The Dreary Steeples

Acts of injustice done
Between the setting and the rising sun
In history lie like bones, each one.

—W. H. AUDEN,
The Ascent of F6, act II, scene v (1936)

When I was a young correspondent in Belfast in the early 1970s, we employed as our babysitter a young Catholic girl whom I'll call Mary. She lived in Andersonstown, at the time one of the most notoriously Republican outer suburbs of the city, where almost every house (including Mary's) provided safe haven for men from the IRA and a secure hiding place for their weapons. Mary and my family kept in touch long after I left Ireland, and eventually at the urging of her parents she left the violence of Belfast behind her and immigrated to Canada, settling in Vancouver.

In due course she married. Her husband, also a Belfast refugee and now also living in British Columbia, was a Protestant. I am calling him Gerald, and he was not just a Protestant, but a truly militant congregant of one of the churches led by the formidably anti-Catholic pastor Ian Paisley. Gerald had lived in that area of Belfast called the Short Strand, arguably the most aggressively loyalist part of town, a recruiting-ground for the Ulster Volunteer Force, a bastion of the Orange Order. He, in short, was everything that Mary would have feared and despised; she was, in Gerald's eyes, the embodiment of papist evil.

Mary and Gerald have been married for forty-five years, are at the time of this writing on the verge of retiring, have raised three children, have an assortment of grandchildren, and are pillars of the Vancouver civil service, Mary involved in childcare, Gerald in highway maintenance. The fact that one is Catholic and the other Protestant barely signifies; nor does the fact that they come from areas of Belfast that remain as unbridgeably hostile today as they were fifty years ago. In Canada, none of this matters. The Irish argument ended the moment they each stepped out onto the Vancouver airport tarmac. Because as both now agree, and as most who think about the political situation of Ireland north and south do as well, the argument in which each were unknowing participants was really all about land.

Gerald and those who thought like him believed, so long as he remained in Belfast, that it was his God-given duty to defend to the death the idea that the land in which he lived would remain eternally under the invigilation of London and would remain steadfastly and loyally part of the United Kingdom. Mary, so often frightened and beleaguered by gunfire as she sheltered in her little council house in Andersonstown, had no such wish, but dreamed instead that one distant day all of the island of Ireland would be peacefully united, as geography suggested it should be, under the equally warm supervision of Dublin.

The argument had little to do with religion, in any meaningful sense. It had very much more to do with such concepts as rule, possession, organization, sovereignty, independence, political union, separation, legal systems—and all of those things that were to be determined by who, in the end, ruled the land on which these two, and all those in Belfast, all in Fermanagh and Tyrone, Armagh and Antrim, Derry and Down, lived out their days. Distance the people from the land

and the argument diminished; return them, and all the old fighting would resume.

It is possible to be too glib about the importance of place to peoples who inhabit lands that are burdened with deep historical fractures. One can, however, point to the relative serenity of at least some of the diasporas around the world, in which those involved were once at each other's throats. The Tutsi and Hutu who fled the genocidal miseries of Rwanda in the mid-1990s seem today to be somewhat more amiably disposed to one another where many now live in Alberta and North Dakota. Serbs and Kosovars in Chicago seem no more hostile to one another today than do those Indians and Pakistanis who live at either end of Devon Avenue on the city's north side. Return each group to east Africa, the Balkans, or the subcontinent and sparks might fly; keep them away from the land that gave birth to their conflict, and an uneasy peace often returns. This is by no means always true, with one noted exception in the next chapter. But all too often enmity subsides with distance and is replaced by the greater notion that all inhabitants of the land surface are simply humans, with more reason to be united than divided.

Such is very clearly the case for Mary and Gerald, who with no family now left in Ireland, seldom go back. Canada is their home, and there are no arguments of such depth and magnitude and longevity there, nor are there ever likely to be. They can just get on with living, and such interest as they have in land now is purely with how best to plant the half-acre plot that surrounds their little retirement house in the foothills of the Rockies. That for them and for now is quite sufficient.

2

The Unholy Land

The futility—not to mention the danger—of Balfour messing himself up with things between Arabs and Jews is beyond castigation. There's bound to be unnecessary trouble there—before long.

—RUDYARD KIPLING,
letter to H. Rider Haggard (1925)

It was on the afternoon of Friday, September 26, 1947, in the former wartime headquarters of the Sperry Gyroscope Company in the Long Island community of Lake Success, that a British government minister—a dedicated socialist, trade unionist, rock climber, and hill walker named Arthur Creech Jones—made an announcement that, though its consequences were unanticipated or, at best, underestimated, would lead to the creation of the most intractable land dispute in the world.

This rotund, bespectacled, mild-mannered, and otherwise somewhat forgettable functionary stood that day to inform the General Assembly of the United Nations—whose delegates were meeting in this insalubrious industrial site while their permanent headquarters was being readied on the East Side of Manhattan—that the United Kingdom would henceforth be ending its quarter century of hapless supervision of the former Ottoman territories of Palestine, and would withdraw all British troops from it by the middle of May 1948.

Arguments over the disposition between Jew and Arab, of these five million acres of hitherto British-mandated Levantine land—with its olive farms and citrus groves and the goatherds' oases and the rock-strewn deserts, with its ancient, ancient cities of unparalleled religious importance, and which lay, mostly fertile and full of promise, between the Mediterranean and the River Jordan, from the frontier with Lebanon in the north down to the Gulf of Aqaba on the Red Sea—would bitterly obsess millions for decades to come, would spur wars and assassinations and acts of terrorism that continue to this day. But after Mr. Jones's singular announcement on that early autumn Friday of 1947, which was the lead story in all the serious British newspapers on Saturday morning, all that really mattered to my family in London was that my father would now be coming home.

As a young soldier he had already had a distinctly trying war. It had begun with some promise: he had commanded a tank squadron in the North African desert in 1943, engaging in a number of modest skirmishes—but then had been unexpectedly summoned back to England for secret training for what turned out to be the D-day invasion, and had then successfully landed with his men on Omaha Beach on the famous June 6 start of Operation Overlord. His luck ran out three days later, however, when he stumbled into an ambush in the deep lanes of the Norman countryside, and was captured and sent eventually to an officers' POW camp, Oflag 79, in the Saxon city of Braunschweig, where he spent the remaining months of the war. When in March 1945 he was liberated by American soldiers he was sent home to London, only to be confronted by two surprises: first, that the family house had been demolished by a V-2 flying bomb, and second, that I had been born.

The War Office was a body little given to sentimentality; and in the pitiless ways of the times, my father was promptly ordered away from London and his ruined house and his newborn child, was placed onto a troopship, and was sent off by way of Gibraltar,

Malta, and Cyprus to the Levantine port of Haifa, in Palestine. There he would remain for two further uncomfortable and dispiriting years, until shortly after Mr. Jones's announcement to the United Nations when, like ten thousand others, he was given his marching orders and told that at last, his duties now all faithfully discharged, he could go home, for good. He'd be given a slim metal commemoration bar to add to the ribbon of his single campaign medal, he would be slathered with thanks, formally demobilized, and allowed to return to life as a civilian.

My father's posting was to a territory suffering through the second of its three most recent iterations. The first had commenced early in the sixteenth century; starting then, and for the next four centuries, this vast acreage—5.5 million acres in full extent, or to use the old Turkish areal unit still employed in Israel, 22 million *donum**—had been administered as

---

* In Ottoman times the *donum*, like the Greek *stremma*, was a somewhat flexible measure, initially based on the amount of land that could be ploughed by a team of oxen in a single day. It was later amended to reflect a square of forty paces each side, but paces in Iraq turned out to be very different from those in Palestine. In recent years the *donum* has been redefined in metric terms and is generally accepted to be one tenth of the area of a hectare, or 1,000 square meters.

part of the Ottoman Vilayet of Syria. Since 1516, when the Ottomans had displaced the Egyptian Mamluks, it had been dreamily ruled by a Turkish *wali* who sat on a golden throne in Damascus, and had as his notional deputy a French-speaking Turk who ran (though he answered directly to the viziers of the Sublime Porte back in Constantinople) what was called the Mutesarrifate of Jerusalem. Up until the mid-nineteenth century these Holy Lands—though substantially peopled by Jews at various periods in antiquity—were almost wholly Arab. In 1850 perhaps 4 percent were Jews, most of them deeply religious scholars buried deep in the souks of the Vilayet's ancient cities. But toward the end of the century these numbers started to climb. They did so for two reasons: a seemingly state-sponsored campaign of pogroms against Jews living in Tsarist Russia, and the formal birth in 1897 of the philosophical notion of Zionism—the dream of stateless Jews around the world to return to the hills of Jerusalem on which the city of David had been built, and re-create a homeland there.

The Ottoman rulers were hardly encouraging, permitting immigration to Palestine, but forbidding newcomers from purchasing any land. By now there was a formalized Ottoman Land Code, with a fully fledged western-style land register and a system that laid down the manner in which the acreages—said to be owned by God or by the Turkish sultan, in common with, for example, the British notion that all British land is owned by the British Crown—could be allocated. The idea of the communal ownership of village land was also floated in the code—a code that outlived the Turks and, somewhat modified, became part of British mandatory rule and later to some extent part of Israeli rule also.

This initial ban on the Jewish purchase of Ottoman acreage was soon cleverly outflanked, so eager were the newcomers to acquire portions of Zion, of their ancient biblically allotted land of Canaan. In a variety of ways, mainly by the employment of proxies, a small gathering of farms and villages managed to pass into Jewish hands—or into the communal hands of Jewish villagers—in the early twentieth century. The population figures started to swell. By 1916 there were some 80,000 Jews in Ottoman Palestine, still less than 10 percent of the total, and the exigencies of the Great War reduced that number again as the fighting in Europe wore on. The Jewish population was still so small back then that few of any non-Jewish persuasion felt threatened, and most residents of prewar Palestine went about their business with a dignified tolerance for one another, Bedouins and bankers, farmers and fellahin all. Most of the rural Jews who in these years seeped in from eastern European shtetls in the various migration waves, or *aliyot,* came to Palestine to set up collective kibbutzim, busying themselves with, as the phrase of the time had it, *making the desert bloom.* While the Turks gazed sleepily on.

All swiftly and dramatically changed once the Allies had begun to celebrate their Great War rout of the Ottoman Turks, which made for the second phase of modern Palestine's history—the British phase in which, toward the end, my father would play his modest and unsung part. It changed because in 1917, already anticipating victory over the Turks, Foreign Secretary Arthur Balfour had issued his famous declaration, pledging support for what the Zionist movement so dearly wanted: the eventual establishment of a homeland for the Jewish people in Palestine.

This was a pledge made not so much out of a particular compassion for the then widely beleaguered Jews, nor an especially keen sympathy for Zionism (at the time, the British ruling class had somewhat more tender feelings—if overromanticized—for the Arabs). Rather, a wily Balfour had offered his pledge, which he made in a well-publicized letter to Lord Rothschild as head of the Zionist movement in Britain, in the twofold hope that influential American Jewry might then support Britain in her prosecution of the war against

Germany and that the likely eventual settlement in
Palestine of large numbers of Jews fleeing the Russian
and now central European pogroms would somehow
help protect the Suez Canal in nearby Egypt, so vital
a communication link with India and the rest of Britain's eastern empire. The Balfour Declaration was in
no sense a decision based on altruism, but, like most of
the more Machiavellian decisions of the British empire,
was an act of policy finely calculated for the playing of
the diplomatic long game.

It had its practical effects almost immediately. Scent-
ing at last the possible realization of the Zionist dream,
those encouraging the migration of Jews to their long
denied gathering grounds in Palestine redoubled their
efforts, and the pace of settlement started to pick
up. The figures were still limited—but consequential
people, many of them outsiders with money and influence, supported the immigrants, and a trend started to
appear. Jews, or their proxies, began to buy up the
land.

Some believed they had a Talmudic duty to do so.
It was an intensely felt compulsion, almost—one so
strong that rabbinic law had this to say about it:

If one buys a house from a non-Jew in Eretz Israel,
the title deed may be written for him even on the

Sabbath. On the Sabbath!? Is that possible? But as Rava* explained, he may order a non-Jew to write it, even though instructing a non-Jew to do a work prohibited to Jews on the Sabbath is forbidden by rabbinic ordination, the rabbis waived their decree on account of the settlement of Eretz Israel.

So, matters of scripture and doctrine and belief now ran headlong into the creaking machinery of an aging and sagging British Imperium. The Allies' defeat of the Turks meant initially that the British army occupied Palestine, to be succeeded after some months by the installation of a British colonial government that took practical administrative charge at the beginning of July 1920. They came, fully equipped with their customary colonial equipage, to rule under the authority of the League of Nations. Bureaucrats from Surrey and Yorkshire, Scotland and Cornwall—well seasoned with running as district officers and public works officers and directors of marine and tax collectors the lives of foreigners in places from Somaliland to Samoa, Melbourne to Kolkata—would now be in near exclusive charge of this tiny sliver of seaside territory, barely two

---

* A Babylonian rabbi often cited in the Talmud, and renowned for his debating skills.

hundred miles from north to south,* of special significance to Christians and Muslims, and sacred in particular to nearly all of the world's Jews. The place had
become formally subsumed into a junior branch of the
British empire, its official name transmuted into the
British Mandated Territory of Palestine.

London thus also had practical authority over the
Holy City of Jerusalem, and with it the first Christian
government there for a thousand years was installed. At
first the newcomers had ambitions galore for their ability to rule at last from this golden and glorious city—
ambitions that were soon to prove themselves tragically
out of sync with the febrile passions of the time, passions that started to show themselves forcefully when
European Jews began to arrive in the thousands to live
among an Arab population not unnaturally surprised
and dismayed by what they saw as the sudden usurpation of their territory.

For a while, the Britons sought to burnish and cherish the loveliness of their ancient possession. With the

---

* A United Nations friend with whom I was staying in Damascus once suggested to me that we go waterskiing in Haifa.
The journey—across the tank-infested Golan Heights, and then
passing through the substantial border fences into Israel—took
less than two hours; after a lunchtime splash in the Mediterranean, we were back home in Syria in ample time for tea.

Temple Mount, the Al-Aqsa Mosque, the Church of the Holy Sepulchre, the Western Wall all now under British administration, the new rulers entertained a host of naïve dreams. Idealists among them hoped that Arab and Jew—who were in British eyes not too terribly different a people, after all—might in time learn to get along and form together a Palestinian governing class. The fantasy held that educated and dignified Arab gentlefolk, plucked from the dunes where they had long been schooled in the same kind of hierarchy and aristocracy that the British then so liked and admired, would mix and mingle with urban intellectual exiles from Warsaw, Berlin, and Kiev, would furnish their own libraries with books from Jewish-run bookshops, would eat cream cakes in Viennese bakeries, would attend concerts given by Bronislaw Huberman's famous Palestine Orchestra, where they would marvel at the masterly grace of Arturo Toscanini's conducting and at the number of priceless Stradivarius and Guarnerius instruments that the exiles had brought down with them. Jerusalem, both Promised and so promising, would in time be restored and made properly divine once again, and at last under properly civilized guidance and rule.

But it was a dream that all turned soon to ashes, acid, and *amertume*. Instead of harmony and propriety there was fighting and revolt, insurgency and ugliness. And all the distemper and disobliging behavior came about, essentially, over the vexed questions relating to the possession of this newly available tract of Levantine land. Just as in Ireland, here in Palestine, land was all.

The Ottoman rulers for most of their four centuries *in situ* had as a matter of strict policy initially prohibited the sale to all foreigners of any of the five million acres of either their Syrian Vilayet or the Jerusalem Mutesarrifate. Toward the end of the nineteenth century they somewhat relaxed this ban, although Jews— many of those few who came before Balfour spoke did so from Tsarist Russia, which was generally hostile to the Ottoman court—were still not allowed to buy. But when after 1919 the British assumed power, they came with a raft of new regulations, issued mainly to protect the fellahin, the members of Arab rural peasantry. No one, said the imperial rulers, could now buy more than sixty acres of land, and they could not spend more than three thousand Palestine pounds in doing so, without government approval.

It was a valiant attempt to control matters, one might say—one measure among many. But the incoming Jews soon found ways to circumvent such regulation, and a wave of purchases soon began on a heroic scale. Many were purchases of land not made by individual Jews, but by wealthy outsiders, by sympathizers and supporters, and by enormous new funds that had been set up specifically to buy land and encourage immigration and settled settlement. The largest and best known of these funds were Baron Edmond de Rothschild's Palestine Jewish Colonization Association (which bought 125,000 acres), the Palestine Land Development Company, and, most significant of all, then and still, the Jewish National Fund.

The eventual ability of well-financed Jews to buy up large tracts of territory in Palestine—many tracts hitherto owned by Lebanese and Syrians and others beyond the Mandate's borders—led swiftly to the creation of a sizable class of landless Arabs workers—men whom the Jews would not then employ, preferring as laborers those whom they felt they could trust. Separation of the two societies, Arab and Jew, began in earnest around then. As matters worsened, Arab riots, revolts, and strikes—staged, called, and organized to protest what was seen as a stealthy expropriation of their traditional lands, a wholesale usurpation of fields they had tilled and on which their goats had wandered for millennia—became a common feature of the period. Intercommunal strife veered out of control on occasion, most especially in 1933 and 1936, and it spread into cities and in manners too many to describe or evaluate.

The British made countless attempts, measures drawn from the vast armory of colonial control elsewhere—royal commissions, judicial inquiries, orders in council, regulations, states of emergency, declarations of martial law—to still the trouble and stem the tide of inrushing Jews. For a while in the 1930s, when there were no more than 200,000 Jews in place, and still four times that many Arabs, whose birthrate (thanks in no small part to improved colonial medical care) was suddenly increasing to levels the British thought might stabilize matters; the demographic mix seemed controllable, manageable, suitable, and appropriate. But then came Adolf Hitler—and the purges in Germany, *Kristallnacht*, the Wannsee Conference, the Final Solution, the Holocaust—and the floodgates opened wide and forever, and the tide roared unceasingly inward.

Though British soldiers, my father included, were ordered to turn the scores of rickety refugee boats away from the ports of Haifa, Hadera, Ashdod, Ashkelon, and Yafo, and to send the frantic dispossessed European Jews off to internment camps in Cyprus and even to faraway Mauritius, the game was evidently up. The British dream was over; the hopes of civility and stability were dashed; the time had come to redeem the promise that Balfour had made thirty years before. Hence Arthur Creech Jones, and his speech in Lake Success.

By the autumn of 1947 the British government had realized what my father and his weary soldier-colleagues on the hot and hard Palestinian ground had already come to know well: the violent passions felt by thousands of these incoming Jews, demanding of a seemingly dithering international community that they should be granted title to their long promised sanctuary homeland in Palestine—a passion that manifested itself in ever greater violence directed at Arabs and those foreign soldiers whose job was to keep the peace—created a situation that was militarily and politically untenable and unsustainable. More than one hundred British soldiers had already been killed by bombs, shootings, and hangings dealt out by the Irgun, perhaps the toughest of the guerrilla groups behind the growing Jewish insurgency. It was abundantly clear that unless Palestine was handed over to the Jewish people, wholesale, and soon, matters would only get worse, becoming a full-scale civil war, spiraling hopelessly out of control.

Accordingly, and understood today by parsing the

circumlocutions of Mr. Jones's U.N. address, it was eventually decided by Downing Street and then formally announced to the world that September day in 1947, that Britain would cut and run, would wash its hands of this troubling corner of its now fast-crumbling empire. The vexed territory would now be formally abandoned, handed over at last to Palestinians and Jews to sort out on their own.

As if to make sure the British truly did get out, the Irgun staged a spectacularly cruel massacre in a village outside Jerusalem, Deir Yassin, just five weeks before the departure deadline. Irgun irregulars came into the tiny community during the night of April 9, 1948, and in the battle that followed at least 120 villagers died, by some accounts 255. There were mutilations, rapes, savagery, and slaughter, and a lasting certainty that whatever blandishments and nostrums might be offered at the departure ceremonies due thirty days later, the new state of Israel would get off to an inconsolably tragic start.

And so it was, that on May 14 the Union Jacks were all struck from public buildings, marines played farewell dirges to the inhabitants of Mandatory Palestine— Jew, Arab, and gentile all the same—and the final detachment of tropical-kitted and mightily relieved infantrymen, under the watchful eyes of a cluster of army generals, boarded their landing craft and bobbed out into the Mediterranean Sea.

The high commissioner of Mandatory Palestine, General Sir Alan Cunningham, who had been based in the badly damaged King David Hotel in Jerusalem, left on a Royal Air Force Dakota once his men had gone. On telegraphed instructions from London he had ordered the officials and the generals to leave this melancholy rump of the empire, announcing as he did so that once he himself had quitted the scene, the territory was to be officially transformed from a mandated possession into a sovereign and independent state, to be named what it is still named today: the Medinat, the State, of Israel.* It would be run under the prime ministership of a Polish refugee and ardent Zionist named David Ben-Gurion, and from his own headquarters in the newly built seaside city of Tel Aviv. Tel Aviv–Yafo,

---

* Before the formal establishment of the *State* of Israel, Zionists used the term *Eretz* Israel or the *Land* of Israel, a southern Levantine entity of a traditionally indeterminate size and shape, its dimensions varying according to numberless references in biblical texts, and representing more the Zionist *dream* of a Jewish state rather than any embordered practicalities. In the quarter-century existence of the Mandate, supporters would call the territory Palestine-EY, meaning Palestine-Eretz Ysrael, conflating their dream with the political realities of the day into a single quasi-national name.

more formally—the onetime location of the ancient desert city of Jaffa, built on land bought from the Bedouin in 1909.★

The declaration that Ben-Gurion then made of the establishment of the new state contained nineteen paragraphs. Probably the most significant of them came halfway down the page, stating:

> The State of Israel will be open for Jewish immigration and for the ingathering of the Exiles. It will foster the development of the country for the benefit of all its inhabitants. It will be based on freedom, justice and peace as envisaged by the prophets of Israel; it will ensure complete equality of social and political rights to all its inhabitants irrespective of religion, race or sex; it will guarantee freedom of religion, conscience, language, education and culture; it will safeguard the Holy Places of all religions; and it will be faithful to the principles of the Charter of the United Nations.

★ To circumvent the Ottoman ban on Jewish immigrants buying land, the first plots for the new city were bought by a Dutch banker and prominent Zionist named Jacobus Kann. His membership in the Dutch aristocracy spared him from some of the initial savagery of the Nazi occupation of the Netherlands, but he and his wife were eventually moved to Theresienstadt in northern Czechoslovakia, where he died in 1944.

The Palestinians have named the day that Israel came formally into being, May 14, 1948, as Al Nakba, the Catastrophe. They scoff, almost all of them, at just about every word of Ben-Gurion's declaration— equality, pah!, freedom, pah!, faithful to principles, pah! They decry, almost all of them, what they see even at this remove as the savagery and rapine and outright theft of their land that took place during the first few months of the existence of the state. The Deir Yassin massacre, they say, was only a small part of the violence that attended the birth of Israel. In tens of thousands, crowds of defeated, dejected, and demoralized Palestinians began to leave, and to become for the time being refugees, displaced people, heading in caravans and cars and camel trains west to Gaza or east to Jordan or else to safer places farther afield. By 1949, some 720,000 had either fled or been expelled. This set the stage for the eventual population of Israel to be overwhelmingly Jewish: of the current 8.7 million inhabitants, 6.5 million are Jews, almost 1.8 million—a fifth of the total—are Arabs.

The Jewish side of the land argument is, quite naturally, the polar opposite of that of the Palestinians. Zionists will argue that what took place during those early days—and in all of the subsequent battlings, including, most notoriously the 1967 war, after which Israel annexed the West Bank of the River Jordan and the seaside Gaza Strip—reflected a reasonable and graduated response to the wholesale policy of existential attacks from disaffected Arabs. The two sides have been arguing ever since, and despite occasional moments of optimism show no real signs of forgiveness or real reconciliation.

And the land remains at the heart of it. By far the greater part of Israel's land is effectively now owned by the state, held in trust for the Jewish people as the government insists; it is leased to those who live there. The huge tracts originally acquired by the Jewish National Fund are administered by the Israel Land Authority—which grants leases wholesale to Jewish Israelis, and with some restrictions that are said to be for security reasons, to Israeli Arabs. There is a small amount of privately held land still, and this may be sold to whoever has the requisite money, be they Jew, Muslim, or gentile.

Much of what occurred during the latter months of 1948 has long since vanished in the rearview mirror. The arguments that dominate today's headlines, all of them the disputatious stepchildren of those early disagreements, make for a dire litany of daily aggravations: the construction of new Jewish homes on land that much of the world regards as illicitly seized, the

bulldozing of Palestinian houses, the firing toward Israeli civilian houses of rocket barrages from Palestinian
strongholds, the construction of enormous cement walls
that cut off Palestinians' access to places that threaten
Israelis' security. All these things and a thousand more
serve to inflame passions on both sides, and to keep
them inflamed for what seems likely to be an eternity.

The fundamental argument that rages within today's Israel concerns the ownership of the land. It is

Construction began in 2000 of a four-hundred-mile long
"separation barrier," consisting of cement walls or barbed-wire
entanglements, surrounding the entirety of the West Bank. The
Israeli government says it protects the country from Palestinian
violence.

an argument that pits Palestinian against Jew in much the same way as the very same question pits Catholic against Protestant in Northern Ireland. Except that there is a signal difference. As Mary and Gerald in Vancouver have so amply demonstrated: Take the Irish out of Ireland, and their dispute, no matter how ancient, somehow evaporates. But take the Levantine out of the Levant, and wretchedly, the argument somehow manages to survive. It clings on like an ineradicable black mold, a dispute that positively refuses to die, no matter how much time elapses, no matter how far from home the parties may have come.

One plausible explanation for the durability of the Middle Eastern land passion is rooted in the sheer antiquity of the dispute. Winston Churchill had famously remarked on the endlessness of the Irish disaccord by noting drily that "once the waters of the Great Flood have ebbed away, so the dreary steeples of Fermanagh and Tyrone would emerge intact once again, since the integrity of their argument is eternal." But in truth, Ireland has little knowledge of the real eternal. What has transpired in Ireland's long and troubled history is as nothing compared with what has been going on in the equally troubled and supposedly very much holier Holy Land. Irish claims to the land of Ireland were challenged most severely by the Cromwellian invasions of the seventeenth century, four hundred years ago. But Israeli and Arab have been at each other's throats about the ownership of the 22 million *donum* of Levantine land for fully *four thousand* years, with both the Palestinian pastoralist and Israeli merchant convinced to the inner deeps of their souls that the land belongs to each, exclusively, and that the Other, whoever that may be, has no right to be there, and by being so is committing trespass, and on an insulting and legendary scale, and so is worth fighting, and to the death.

As to whose land it truly was, has been, and now properly is, none shall ever agree.

Geology, geography, and topography have long determined some goodly portion of the region's fate. Palestine's dunes and fertile plains were on the ancient caravan route between Egypt and Mesopotamia, and like so many of the world's passageways today—the Strait of Hormuz, the Fulda Gap, the Khyber Pass, Suez, the Hudson-Mohawk Gap, the Strait of Malacca—they have long been a place of fierce contest, of permanent watchfulness, and of determined protection. The Israelites, one of the many local Egyptian tribes, passed through and over this passageway in a complex tarantella of expulsion and exile, of settlement (the first time in around 1400 BCE), as well as of massacre and misery, and which they endured for centuries. Through all these dramas and travails, and no matter how far and wide they settled and were dispersed as a result, their traditions and rituals were all sedulously and near miraculously preserved. The ambitions of nineteenth-century Zionism eventually brought them back, to emerge blinking in the sunlight from their various adopted European settlements, and returned them back to their homeland, to this ancient nomads' passageway, and with which they can reasonably claim to be most commonly identified.

The Bible told and mythologized the story—and it had much to say about the land (most especially detailed in the Book of Joshua) that would be tilled and improved by the wanderers. Leviticus reminded them, though, that the surface belonged to Yahweh, and that it "shall not be sold for ever: for the land is mine, and ye are strangers and sojourners with me." Moreover, land management was a topic for biblical admonishment: every fifty years—after forty-nine years of "regular time," as Leviticus chapter 25 verse 9 famously describes it, there would be a Jubilee, a time when the land would be allowed to lie fallow, the Hebrew slaves would be set free, and all the land and the houses in the open country that had been previously sold would revert to their former owners or their heirs.

The English word *jubilee* comes from the Hebrew

word *jubil* (לבוי), which means "trumpet," and which is blown at commencement of the event—a joyous celebration of the land and all that it offered to the people who lived upon it, cared for it, and drew their sustenance from it.

And of all the various claims made upon the five million acres of today's Palestine, this—the biblical notion of *Jubilee*—is to the scholar-Jew the single most alluring element in its justification. It suggests that for thousands of years past the settlers have not merely grown their olives and herded their goats and drawn water from their wells but have cared for and managed and shared land that in any case belongs to God, not to any of humankind. The Arabs see it all as Catastrophe, the Jews as Jubilee, and the argument of such fiercely opposed philosophies continues today, as it has continued for centuries.

And meanwhile the landscape of Palestine itself, drenched though it may be in misery and mayhem, remains entirely neutral, sand and pebble, boulder and cliff, river valley and salt lake, beach and oasis and city street, all of it a child of geology and geography, all placidly accepting its role as, in this one corner of the world, an entity to be fought over endlessly, enduring it all because it alone endures.

3

Death on the Rich Black Earth

I have made a shroud for my friend,
Sad cloth.
She loves, loves blood
This Russian earth.

—ANNA AKHMATOVA,
Poem Without a Hero (1940–1965)

Arguments over land take many forms in many places and have been going on for many centuries. One of the more poignant in recent times involves the violent intersection of a dispute involving vast tracts of territory—millions of broad acres of exceptionally fertile land in central Europe—with the craft of journalism. And in particular with one young British journalist who in the spring of 1933 discovered something terribly wrong, a crime of unfathomable proportions that was then unfolding in the central plains of the Ukraine.

His name was Gareth Jones, and he was born in 1905 in South Wales, the son of a local schoolmaster. He died on the eve of his thirtieth birthday, far away from home in western Manchuria. He had been murdered, shot three times in the head.

It has long been believed that this young, slight, bespectacled, phenomenally talented Welshman was killed by Chinese or Japanese agents working for the NKVD, the Soviet Union's much feared domestic secret police. If this belief is correct, then it is further believed that he was murdered in retaliation for a series of newspaper eyewitness reports that he had written two years before, in which he bluntly accused the Soviet Union of confiscating vast tracts of arable land in the center of Ukraine, and in the process starving millions of people, creating a genocide of near unparalleled ferocity and longevity. Gareth Jones has been named a hero in today's Ukraine, and there are statues in his honor and any number of plaques and memorials, as well as a Hollywood movie.

What made his dramatic and terrible report, which was first published in a London newspaper, the *Eve-*

ning Standard, on Friday, March 31, 1933, of such monumental and lasting importance was its immediate denunciation—but not a repudiation by the Soviet authorities, who one might think had ample reason to refute suggestions that their policies were spawning a widespread famine in their southern provinces. Instead, the rebuttal was issued by Walter Duranty, a journalist for *The New York Times,* a native Briton born in Liverpool and who for the eleven years previously had been based in Moscow. He was a well-connected figure of considerable influence, eminence, and power, and he had already won a Pulitzer Prize for his extensive assessment—in a groundbreaking, multipart series—of the social and political effects of the Bolshevik revolution.

The tragedy that was reported by Gareth Jones—not a whisper of which made it into any of Duranty's initial reporting—turned out to be wholly true, and when fully confirmed, it shocked the world. Ten million peasants had been starved to death, maybe more. The unfolding events of 1932 and 1933 in the Ukraine are regarded by most today—though, notably, not by all—as constituting a classic case of a genocide, and of a genocide far greater in number than any other in properly recorded modern history.

The event is now known in Ukraine and beyond as the Holodomor—murder by starvation. It is now known to have been the result of a fit of Stalinist policy madness born of a tyrannical and ideologically driven desire in Moscow to have the land taken away from the ordinary peasantry of Ukraine, for it to be turned into great collectivized and industrialized farms, which would be established in a way that would be entirely consonant with the Marxist ideal of public owner-ship. The policy instead created a human disaster of unimaginable proportions, and Gareth Jones—though he would soon thereafter pay the ultimate price for its exposure—was the first to reveal it to the world.

His first revelation appeared—not under his own byline, but under that of another Pulitzer Prize win-ner, H. R. Knickerbocker, who had listened to Jones at a press conference he held in Berlin on his return from his surreptitious expedition to the Ukraine—as the lead in the *New York Post* of Wednesday, March 29, 1933. The headline was stark: FAMINE GRIPS RUSSIA, MIL-LIONS DYING . . . SAYS BRITON. The story was picked up by a host of other newspapers the next day, and on the day after that, Friday, Jones was able to publish his own piece in the *Standard*:

All that is best in Russia has disappeared. The main result of the Five-Year Plan has been the tragic ruin of Russian agriculture. This ruin I saw in its grim reality. I tramped through a number of villages in the snow of March.

I saw children with swollen bellies. I slept in peasants' huts, sometimes nine of us in one room. I talked to every peasant I met, and the general conclusion I draw is that the present state of Russian agriculture is already catastrophic, but that in a year's time its condition will have worsened tenfold. If it is grave now, and if millions are dying in the villages, as they are, for I did not visit a single village where many had not died, what will it be like in a month's time? The potatoes left are being counted one by one, but in so many homes the potatoes have long run out. The beet, once used as cattle fodder, may run out in many huts before the new food comes in June, July and August, and many have not even beet.

"Have you potatoes?" I asked. Every peasant I asked nodded negatively with sadness.

"What about your cows?" was my next question. To the Russian peasant the cow means wealth, food and happiness. It is almost the center-point upon which his life gravitates.

"The cattle have nearly all died. How can we feed the cattle when we have only fodder to eat ourselves?"

Walter Duranty's reply appeared on page 13 of *The New York Times*, on the same day, and under the less than reassuring headline RUSSIANS HUNGRY, BUT NOT STARVING. He refers to the Ukraine situation in his 800-word piece, cabled from Moscow in response to Jones's Berlin news conference of a few days before, clearly having been pressed into service by his foreign editor, who must have been stung by the jeremiads in the competing Manhattan papers. Somewhat wearily, Duranty describes Jones as a man "of a keen and active mind" who had "taken the trouble to learn Russian, which he speaks with considerable fluency"; But then he goes on to denounce the Welsh journalist in tones of patrician superciliousness. He felt, he said, that Jones's judgment was "somewhat hasty" and was based on "a forty-mile walk through the villages in the neighborhood of Kharkov and had found conditions sad. I suggested that that was a rather inadequate cross-section of a big country, but nothing should shake his conviction of impending doom."

A photograph of Duranty taken around this time shows him and two contented-looking guests dining comfortably in his Moscow flat, attended by a servant with an uncanny resemblance to the young Trotsky pouring a glass of red wine. The table is set with candles, silverware, and fine napery. The dinner plate in front of the photographer's seat appears loaded with food. In the image the diners may well have been hungry, but most certainly were not starving, nor probably ever had been.

Duranty was prolific, and the articles he wrote during the next few days* kept to the theme of good harvest and prosperity and general success of the growing and now fifteen-year-old Communist juggernaut. SOVIET INDUSTRY SHOWS BIG GAINS, on April 6. SOVIET HAILS GAINS IN SPRING SOWING, April 7. SOWING IN RUSSIA CONTINUES TO GAIN: NEW POLITICAL SECTION OF TRACTOR STATIONS SPURS WORK WITH FOOD, April 10. And then, on April 21: SOVIET SOWN AREA TRIPLE LAST YEAR'S . . . OPTIMISM IS MANIFEST. KARL RADEK SAYS: "IF CROPS ARE GOOD, WE CAN TELL WHOLE WORLD WHERE TO GET OFF."

As winner of the Pulitzer Prize the previous year Duranty enjoyed considerable authority, enabling his criticism of Jones to be widely believed at the time and his evident admiration of Josef Stalin to be accepted as reasoned and reasonable. But only at the time. As

---

* The other story demanding Mr. Duranty's time and attention that April was a big political show trial in Moscow of three Britons, two of whom were sent to jail and the third expelled after being found guilty of spying and sabotage.

Colorful agitprop posters extolled the supposed virtues of the collectivization of Ukrainian agriculture in the 1930s, urging the people to believe the policy would bring prosperity and happiness for all.

the truth about what had actually happened in the Ukraine became ever more clear, and as the truth about the insane savagery of Stalin and his henchmen became ever more apparent, Duranty's reputation began to suffer. Malcolm Muggeridge, who had reported vividly—though anonymously—for the *Manchester Guardian*, declared Duranty to be the "greatest liar" he had ever known, and Joseph Alsop, the celebrated midcentury American syndicated columnist, said that "lying" was Duranty's "stock in trade."

The gathering uproar eventually persuaded *The New York Times* in 1990 to break rank with their long dead correspondent's official reputation: the newspaper denounced him publicly and decried his reporting from Moscow sixty years before as "some of the worst reporting to appear in this newspaper." The storm led to a move to strip Duranty, posthumously, of his Pulitzer. But the Pulitzer board, both in 1990 and then again after a further *Times* appeal in 2003, declined to revoke the award. There was no evidence, the Pulitzer board said, in a masterly piece of pusillanimous evasion—and despite mountains of evidence from former journalistic colleagues, from intelligence agencies, and even from the FBI—that Duranty had known perfectly well that what he was writing was only vaguely concealed Kremlin propaganda, that his deception had been deliberate. Had it been done on purpose, the board said, then the prize would have been withdrawn. Altogether it was an unedifying chapter in the history of American journalism, although it can fairly be said that the *Times* rose to the occasion when the moment demanded, setting in train measures that have ensured—one hopes—that such slovenly reporting would seldom occur again.

What Gareth Jones discovered—and really did discover—during his condescendingly described "forty-mile walk through the villages" was the appalling practical consequence of an act of Kremlin policy that had been hammered out in the spring of 1929, and which was targeted unequivocally at both the peasantry who worked the fields in the country's southern agricultural belt and the land that they owned. The stated intention of this policy was to destroy one specific component of the USSR's peasant class, and to wrest their land away from them wholesale, confiscating it and turning it over, in immense quantities, to the ownership of the state.

Stalin saw to it that this policy—the so-called Five Year Plan, the first of several such economic stimulus programs bent on promoting massive changes to the new country's industrial and agricultural sectors, went into formal practical effect on October 1, 1928. This was some six months before the plan's details had been completed, and so it was as a result both hastily and shoddily contrived, and not surprisingly turned out to be a massive debacle. *The Times* of London reported on July 9, 1929,* that one foolhardy party official, Ser-

* The article appeared on the "Imperial and Foreign News" page under the headline SOVIET INDUSTRIES: FAILURE OF FIVE-YEAR PLAN. The erudition of the paper's readers of the day is

gey Syrtsov, had publicly lambasted the program. In doing so, and by identifying for Stalin the one class of the Soviet people whom he thought to blame for the shambles, he helped to unleash the very tragedy with which the Five Year Plan later became so notoriously associated.

The damnable word that Syrtsov uttered was *kulak*, which in literal translation, and in its Turkish etymology, means "fist." A kulak in the Soviet Communist lexicon was a tight-fisted person—a relatively prosperous man of the countryside who owned a small amount of land and made a modest profit from it. Such parasites, as they were viewed in the eyes of the revolutionaries, and of whom there were millions in the immensely wide grain-growing regions of the southern and eastern Soviet Union, had been identified as the most verminous creatures in the population of the USSR. They were a class of people worthy of utter public contempt,

---

amply demonstrated by some of the other stories that jostled for space, most under headlines needing no explanation for those readers: THE ORAKZAI QUARREL IN THE TIRAH; THE AFGHAN WAR, SALAMIS DISPUTE SETTLED; and, most cryptic of all, THE KONIGSSTUHL OF RHENS, which referred to the removal of a structure made in 1376—and which had once been used by the Elector of the Holy Roman Empire—to a more prominent site on the cliffs overlooking the River Rhine.

just as were the greasiest of landlords or the greediest of capitalists.

The kulaks stood between the might and wisdom of revolutionary doctrine on the one hand and, on the other, those impoverished peasants who worked the land and did not own it and who were in the eyes of Marx, Lenin, Engels—and Stalin—the very salt of the earth, those on whose capable shoulders the whole gigantic engine work of Soviet revolution could be supported. If the Five Year Plan had gotten off to a shaky start, in the view of Sergei Syrtsov then it was principally owing to the fact that the poor and the middle peasantry—the two classes were firmly divided in the Soviet Communist mind—were beginning to show signs of falling under the malign and counterrevolutionary spell of the kulaks. "The middle peasant has turned against us, and has sided with the kulak" was the sentiment most commonly expressed. To rev the plan up and into full-throated action, the kulaks as a class needed to be eliminated, their lands taken away once and for all, the stench of these most evil *bacilli*— that word bandied about by most of the attendees— needed to be eradicated, for all time.

The kulaks had long been hated by the revolutionary purists. A memo from Lenin himself came to light recently in an archive at the Library of Congress in Washington, D.C. It consisted of an order, sent to his faithful Soviet deputies superintending Penza, a region—a *volost*—four hundred miles southeast of Moscow.

SEND TO PENZA
TO COMRADES KURAEV, BOSH, MINKIN AND
OTHER PENZA COMMUNISTS

Comrades! The revolt by the five kulak volosts must be suppressed without mercy. The interest of the entire revolution demands this, because we have now before us our final decisive battle "with the kulaks." We need to set an example.

1. *You need to hang (hang without fail, so that the public sees) at least 100 notorious kulaks, the rich, and the bloodsuckers.*
2. *Publish their names.*
3. *Take away all of their grain.*
4. *Execute the hostages—in accordance with yesterday's telegram.*

This needs to be accomplished in such a way, that people for hundreds of miles around will see, tremble, know and scream out: let's choke and strangle those blood-sucking kulaks.

>*Telegraph us acknowledging receipt and execution of this.*
>
>*Yours, Lenin*
>*P.S. Use your toughest people for this.*

Once Sergei Syrtsov's message, so similar in sentiment to Lenin's of a decade before—and perhaps as an apposite memorial to him, now dead for four years—began to echo its way down from corridors of the Kremlin and thence down to the Party apparatchiks on the ground in the Ukraine and beyond, so the plan began to gather muscle and energy, and the misery began to unfold. By the summer of 1930, the first signs were beginning to show themselves. Kulaks were being rounded up and sent away—and the confiscations, the grain hoarding, the executions, and the hunger began. These policies would gather strength in 1932 and by early 1933, when Gareth Jones, acting on a tip that strange and terrible things were happening in the Ukraine and in the countryside south of Moscow, made his first visit and managed to tell the world of an unfolding tragedy of quite dreadful proportions.

In essence, what had happened since the start of the attacks on the kulaks and the true beginnings of the famine proper had borne out Syrtsov's initial apprehensions: the kulaks and the poorer peasantry had indeed rebelled—both at the first attempts to collectivize their farms, and also at the huge amounts of grain that the farmers were being ordered to send out of the region, to feed the immense and growing armies of the newly industrializing country. The kulaks didn't want to give up their very modest but somewhat secure lifestyles, and the region as a whole didn't want to let the food it produced for its own consumption be spirited away to distant factories. By all means, the kulaks said to the Party commissars, take the surplus, as had always been the case; but leave enough for the locals to be able to eat.

At first the collectivization machine worked fairly effectively, if brutally. Local Party chiefs saw to it that landowners, however puny their holdings, were turfed off their properties, their farm equipment was confiscated, fences were torn down, and tractors and harvesters were moved to gigantic newly built garages. "All boundary lines separating the land allotments," came the order from Moscow, "are to be eliminated and all fields are to be combined in a single landmass." The giant new *kolkhozes*—it was a new word, coined specifically to denote a collective farm—began to sprout all through the spreading reaches of the nation's grain-producing regions; but the *agrogoroda*—another newfangled word meant to signify the enormous agricultural cities where the displaced peasants would be housed—never did take hold, and such new population centers as were built were known, less pretentiously, simply as kolkhoz settlements

That was how it was all supposed to happen—and indeed, for a short while, the success rate was prodigious: the number of families absorbed into collectivized farms reached eleven million in the first two months of 1930. Stalin himself started to fret that the rate of change was too fast for the peasantry: famously, he wrote an essay in March 1930 for *Pravda* saying that he was "dizzy with success" and that maybe some of his Party shock troops and those of the twenty-five thousand hastily trained proselytizers sent down from the capital might ease off for a while, so as not to unduly annoy the farming community.

He was wasting his breath. The locals had already become angry, rebellious even. By the end of 1930, the success and extent of a simmering mutiny had wholly infuriated the Moscow leadership, and despite his initial concerns, Stalin most especially. The much vaunted Five Year Plan was seeming to sputter, again. The Politburo's public relations machine, normally so skilled at producing the kind of propaganda that Walter Duranty would be happy to write for his audience back in New York, was beginning to sense that they were losing control of the story. Factory workers needed to eat. Socialist principles needed to be firmly applied to the kulaks' smallholdings. The twenty million small family farms that existed in the Soviet Union in 1929 now needed to be converted, vigorously, into some two hundred thousand kolkhozes—a feat of territorial prestidigitation that could be achieved only through coercion, threats, or brute force. Moscow's orders now needed to be obeyed. And so Moscow felt obliged to act, decisively, to bring the wavering peasantry into line.

This it started to do in the summer of 1932. Communist agents fanned out into the fields, demanding disciplined acceptance of the New Economic Policy (as the Five Year Plan was officially called), that kulaks give up their holdings forthwith and obey instructions with immediate effect, such that all crops produced in the new farms henceforward be sent to the state warehouses, and that not one single ear of grain be kept behind for the peasants' own use. They might grow the food. But they had no right to use it, either to feed themselves or to nourish their families. The land was no longer theirs but belonged unequivocally to the state. Anyone disagreeing with such precepts would be arrested, deported to the gulags, or shot on the spot.

And if there was suffering caused by this new policy, it was by no means the fault of the new policies. Rather, to quote Stalin's words, it had come about because "the old government, the landlords and capitalists, have left us a heritage of such browbeaten peoples . . . these peoples were doomed to incredible suffering."

The suffering began swiftly. The trials, the shootings, the miseries of the de-kulakization policy, the attacks on churches (which were so central to village life), the deportations, the summary executions—all these were horrific enough. But what remains most haunting was the growing hunger of a people who were being completely denied their food, and who spent their dying winter days searching frantically for something to feed their crying children, let alone themselves. But there was nothing: warehouses filled with grain destined for the factories were guarded by lantern-jawed militiamen, and anyone caught stealing grain or hoarding even the slightest amount was exposed for sabotage or counterrevolutionary activity and hauled off, most probably to be shot. The villagers were silent, cowed, and starving.

The best account of the Holodomor remains *The Harvest of Sorrow*, written in 1986—when, admittedly, the archives had not been as fully opened as more recently—by the British historian Robert Conquest. He quotes a Ukrainian villager's description of the searches for hidden food, carried out by activist brigades, much like the Red Guards of Mao's China. These brigades, writes Conquest,

> consisted of the following persons: one member from the presidium of the village soviet, or simply any member for the village soviet; two or three Komsomols; one Communist; and the local schoolteacher. Sometimes the head or another

A slow and painful death from hunger was the fate for millions of Ukrainians during the Holodomor, a famine that is still widely regarded as a genocide specifically ordered by Stalin.

member from the cooperative administration was included, and, during summer vacations, several students.

Every brigade had a so-called specialist for searching out grain. He was equipped with a long iron crowbar with which he probed for hidden grain.

The brigade went from house to house. At first they entered homes and asked, "How much grain have you got for the government?" "I haven't any. If you don't believe me search for yourselves" was the usual laconic answer.

And so the "search" began. They searched in the house, in the attic, shed, pantry and cellar. Then they went outside and searched in the barn, pig pen, granary and straw pile. They measured the oven and calculated if it was large enough to hold hidden grain behind the brickwork. They broke beams in the attic, pounded on the floor of the house, tramped the whole yard and garden. If they found a suspicious-looking spot, in went the crowbar. In 1931 there were still a few instances of hidden grain being discovered, usually about 100 pounds, sometimes 200. In 1932 however, there was none. The most that could be found was about ten or twenty pounds kept for chicken feed. Even this "surplus" was taken away.

But by 1932 there was nothing left to find, no surplus, nothing to hide—and so in their thousands, people began to starve. Limbs withered. Stomachs began to swell with kwashiorkor. Heads began to loll. People staggered and fell by the roadside, dying in full view even as others limped painfully by. Even the very appearance of imminent starvation caused the brigades to take interest, not pity—the assumption being that if you were starving but alive you must have a stash of illegal food somewhere. One brigade member "after searching the house of one peasant who had failed to swell up finally found a small bag of flour mixed with ground bark and leaves, which he then poured into the village pond."

It is important to remember that the central plank of Stalin's draconian policy was the collectivization of the land—a policy that was at first voluntary but had later to be enforced because of the vehement opposition of the kulaks. That being said, it is impossible to minimize the tragedy that resulted—whether it was the state-directed annihilation of the kulak class, the catastrophic failure of the policy itself, or the failure of Soviet communism as a whole, fifty years later in the nation's history. Nor is it possible to minimize the nature of the starvation, as another activist, tortured by remorse and quoted by Robert Conquest, recalled:

With the rest of my generation I firmly believed that the ends justified the means. Our great goal was the universal triumph of Communism, and for the sake of that goal everything was permissible—to lie, to steal, to destroy hundreds of thousands or even millions of people, all those who were hindering our work or could hinder it, everyone who stood in the way. And to hesitate or doubt about all this was to give in to "intellectual squeamishness" and "stupid liberalism," the attributes of people who could not see the forest for the trees.

In the terrible spring of 1933, I saw people dying from hunger. I saw women and children with distended bellies, turning blue but still breathing but with vacant, lifeless eyes. And corpses—corpses in ragged sheepskin coats and cheap felt boots; corpses in peasant huts, in the melting snow of the old Vologda, under the bridges of Kharkov. I saw all this and did not go out of my mind or commit suicide. Nor did I curse those who sent me out to take away the peasants' grain in the winter, and in the spring to persuade their barely-walking, skeleton-thin or sickly-swollen people to go into the fields in order to fulfil the Bolshevik sowing plan in shock worker style.

Nor did I lose my faith. As before, I believed because I wanted to believe.

The cruelest regulation that Stalin enacted and saw brutally enforced has come to be known as "the law of five ears of wheat." It was enacted in August 1932, and it held that any gleaning—any removal of leftover grain from collective-farm fields that had been harvested— was henceforward punishable by imprisonment for at least ten years, or by death, with executions performed on the spot. Even children caught picking leftover ears of grain would be taken away, and either deported or shot. Those Ukrainians left behind were terrified; those who survived, and who live today, remain suffused with anger and bitterness toward the Russians.

I spent some weeks in the searing summer of 2019 traveling down the Dnieper River valley, south of the Ukrainian capital of Kyiv, into a region as much the breadbasket of southeastern Europe as Kansas and Nebraska are the granaries of the United States, and Saskatchewan and Manitoba of Canada. The landscape is rolling and lush and very, very fertile. High hedges or lines of tall poplars wall off fields that in the high summer are butter yellow with acres of wheat and or deep green with corn, brilliant yellow with rapeseed, or paler green with soybeans or broccoli, potatoes, barley, beet, or sunflowers. Some eight million acres of Ukraine's land is rich with thick, black soil, soil so obviously fertile that it almost looks good enough to eat without any tiresome need to pass vegetables through it. The endless fields stretch to the horizon—more human-scale here than in the American prairies, since there are hills and streams and patches of forest rich with wild boar, pigeons, and partridges. The farmers here are as happy a breed as farmers are ever likely to be, proud of their land, thankful for the riches that are the bounty of this very special portion of the Earth.

And they give their thanks in the domed Orthodox churches that can be seen in every village across the countryside, and in which the villagers pray with enthusiasm each Sunday, and often more frequently than that. They prayed and gave thanks in churches in all the villages that I visited: small places like Rosava and Pustovity, and in larger towns like Mironovka and Vladyslavka. And they paused outside too, to spend moments in the cemeteries, where the victims of the Holodomor are buried. Each memorial has, fashioned from a piece of bent iron wire, a gathering of ears of wheat tied with an iron ribbon, in commemoration.

Nowadays the graves of the Holodomor dead have been joined by those, now neatly trimmed, of young soldiers from the east of the country, who have been fighting in vain against the Russian invasion, and whose parents or brothers and sisters and wives pray that one day the Russians will go away, will leave Crimea, will leave Ukraine in the kind of peace these people seem never to have enjoyed, and which most believe they never will.

For the grim truth is that Russia's intentions toward Ukraine's landscape are near eternal, and are rooted in history, language, religion, and emotional connections, all of which are as powerfully magnetic as they are difficult either to define or to quantify. As long ago as the ninth century, the Kievian Rus was the spiritual center of what would become the Russian empire, the birthplace of the Orthodox church, the location of the vast wealth of the *chernozem* soils whose crops were destined to feed the vast neighbor nation to the north. The evils that Stalin perpetrated against the Ukrainian people will never be forgotten from the Belarus border to the Black Sea, from the Carpathians to the Urals, and the Ukrainians retain an anger and a bitterness that no emollient words from Moscow will ever manage to soothe. This land is their land, all Ukrainians firmly and ineluctably believe.

But Russia needs the bounty of this land too, or

firmly believes that it does—and Russia is so vast and so needy, and its people so reliant on the gifts with which Ukraine's geography and geology have for so long bestowed upon them, so they believe, that they will probably never relax their claims. And this will be so, no matter how many Ukrainians turn up in church each Sunday and bow their heads in reverence before the wire-framed ears of wheat in the country churchyards, and no matter how long, chiseled into the great black marble slabs in the immense new Kyiv Holodomor Museum, are the meticulously maintained lists of the names of the millions who died on these lands, victims of one of the twentieth century's most appalling genocides.

That the land itself survives, and is now so fertile, is a savage irony, considering how many millions who once lived on it and farmed it with such evident and placid contentment died so horribly. They survive today only in the popular memory, their names incised onto marble memorials, or perhaps mentioned in now-yellowing newspaper articles written by heroic young journalists like Gareth Jones, that teacher's son from the south of Wales, who told his stories when he was not yet thirty years old. And with the victims now buried in the rich and fertile black soil of the land on which they starved.

4

Concentration and Confiscation

HERE WE ADMIT A WRONG
—Carved into the sandstone of the
U.S. National Park Service Memorial to
Japanese-American Patriotism, Washington,
D.C., from a formal apology by
President Ronald Reagan (1988)

It was a Monday morning in late April 1944, and an unseasonably chilly and dust-laden wind was howling down the Idaho plains from the Sawtooth Mountains. It was then, so far as we can reconstruct, that a thirty-one-year-old former strawberry farmer named Akira Aramaki was shaken awake in his prison barracks bed by a party of soldiers. Somewhat unusually, the men were smiling. They didn't say why, but they told him to walk, and preferably at the double, across to the camp office—fully half a mile away, the camp being so large and spread out—to collect an envelope.

Mr. Aramaki suspected what was coming. He had been waiting for this moment for the past month, ever since rumors began to swirl in the late winter, when he and the few thousand other Americans of Japanese origin who were cloistered in the huts around him had been told that some of them no longer deemed "a threat to national security" were going to be freed from their two years of imprisonment in this vast concentration camp complex known as Minidoka. There were nine other such camps spread out across the western states, and some 120,000 others like Aramaki, his wife, and their newborn son were also about to be given their long delayed freedom. So he walked steadily, but then, with gathering excitement, he started to trot and finally to run between the barrack blocks up to the front door of the administration building. A cluster of others— some of whom he knew; but with 8000 prisoners in the camp, he was only on nodding terms with most—were already there, lined up waiting to be called.

It all happened swiftly enough. "Aramaki!" yelled a civilian, an official of the War Relocation Authority, which ran the government camp. He handed the young farmer a large brown manila envelope and then shook his hand, grinning as he did so.

As expected, the envelope contained three items. There was a certificate, under the embossed eagle-and-arrows insignia of the U.S. army and signed by a two-star general named Lewis, attesting to the fact that Aramaki was now a free man and should be accorded all courtesies as might be extended to any other American citizen. There was a clutch of five crisp five-dollar bills, held together with a paper clip. And there was an official army movement voucher, a ticket that was valid for use during the following seven days; it would permit him unrestricted one-way travel by bus or train from this remote spot in the Snake River plains of southern Idaho back to his hometown—and his long neglected strawberry fields—some six hundred fifty road miles westward to Bellevue, in Washington State.

There was a clutch of buses outside the barbed-wire fences. If Aramaki would go back to his barracks and collect his few belongings, said the guards, one of the buses would take him and other freed prisoners to the Union Pacific railroad depot in Minidoka town, fifteen miles away. His wife and child would have to remain behind, the guards said, without further explanation, other than to tell him it was regulation, and he should not worry unduly. Likewise, his mother, similarly detained, would stay for a while longer, also for unexplained reasons.

At the train station they wouldn't be able to board the magnificent Streamliner service, which sped well-heeled passengers from Chicago out to Portland in Oregon, since that train passed by each early evening without stopping. But there was a local train, the Portland Rose, which made a brief stop at Minidoka halt each morning at 11:15. It would be a trying journey, especially in spring, as the high passes over the Rockies were often blocked and his train might well have to wait for a plough to clear the way ahead. But if all went well they'd be in Boise, the state capital, by lunchtime, would cross into the Pacific Time Zone and make a ten-minute dinnertime stop in Huntington, Oregon, to take on water. Then, by following the course of the Snake River the train would make Pendleton by midnight, see dawn at the town of Hood River, reach Portland by breakfast, and finally draw up with a hiss of steam and a squeal of brakes at Seattle's Union Station terminus at two in the afternoon. A taxi—if anyone at the station would take him; he knew well of the likely hostility felt toward Japanese this late in the war—should bring him to the farm an hour or so later.

But Akira Aramaki didn't care how long it took. He was now at last heading home, and if not with all of his family members, he would at least be reunited with the land that for nearly thirty years he and his parents had so patiently tilled and cared for. It was a ten-acre plot which produced fruit crops of such abundance that they had helped make the small town of Bellevue the strawberry capital, so-called, of the United States.

But what happened next was not quite as he had anticipated. For while he had been absent, unjustly imprisoned by the American government, someone else had seemingly taken his land. Aramaki found he had, or so he thought, been dispossessed.

The patriarch of the Aramaki family, Hikotara, had arrived in the United States in the late autumn of 1904. He had come from Kumamoto, in southern Japan— a city known today for its tiny sweet-tasting oysters, but which at the turn of the last century was still mired in the traditions—of shoguns, samurai, steadfast isolation—that had been so rudely overthrown by the arrival of Admiral Matthew Perry and the American "black ships" of fifty years before. As a result it was a community somewhat insulated from the modernizing zeal that was then roiling the country. Japan was also not in a good place in 1904, the year Hikotara decided to leave. The Russo-Japanese War was in full force, and such treasure as was in the government coffers was being spent on the conscripted soldiery, not the coun-

try, which was as a result widely impoverished. Out-
lying cities and prefectures were very much harder hit
than the larger industrial and commercial centers like
Tokyo, Yokohama, and Osaka. Impoverished men living in the more remote parts of the country—in places
just like Kumamoto, where they were endlessly fearful
of being press-ganged into the army for service on the
Russian front—heard tell of better working conditions
in America. Recruiters were also abroad in the countryside, telling of high-paying jobs in this land of promise.
And so in the hundreds and then the thousands, men
began to venture across the Pacific Ocean in search of
improvable lives. Hikotara Aramaki was one such: he
was moderately well born, belonging, his family said
later, to the samurai class, and found it fairly easy to
rustle up the fare for a long sea journey from Yokohama
aboard a steamship bound for the only destination he
could find: the Canadian port of Vancouver.

He arrived there in the chill of December—and
soon thereafter he set about walking—for reasons
unexplained—down to the United States border and
then, once across, to the nearby town of Bellingham.
He and another Japanese he had encountered en route
decided to stay for a while, opting not to head farther
south for the city center of Seattle, largely because
they had been told of a virulent anti-Asianism that

had lately sprung up there. Instead, they both took jobs working for the railroad, with Hikotara, perhaps because of his commanding samurai manners, rising swiftly to become a section foreman. He was much liked by his colleagues down the line: James J. Hill, they called him, alluding to the big railroad boss of the Pacific northwest for whom, ultimately, they all now worked. He soon moved on to work a new branch of the line, just built across on the eastern side of Lake Washington, in a much more raw and wild stretch of countryside. The line served a cluster of villages, and a fair number of other Japanese migrants were already employed by the company as section hands or engine watchmen, performing work that white men didn't care for and at wages that were no more than a pittance. But these men, many of them having come from farming communities in Japan, immediately saw the potential for performing agriculture—both in the logged lands that lay, partly cleared, on either side of the new railroad tracks, as well as in those razed areas where logging companies had more fully cleared the forests to provide timber for the building of Seattle's housing across the lake.

Hikotara Aramaki radiated authority to contemplate leaving the business of railroading and setting out on his own. Once he had come to understand the way things worked in this still somewhat undeveloped, pioneer-suitable countryside, he and some of his similarly ambitious fellow migrants suggested to a gathering of the local landowners—most of them lumber companies—a scheme that they thought could benefit both sides. They would clear the land of tree stumps—a miserable task that, once again, white men were loath to perform—and in exchange would win the right to cultivate the cleared acreage, growing fruits and vegetables on it. They would then rent this land for five-year terms and grow what the rich, damp, and amply mulched earth suggested would be best suited: lettuce, celery, cabbage—and strawberries.

The arrangement worked exceedingly well. Hikotara Aramaki took to land clearing with a vengeance. A single acre might sport thirty huge stumps, and these had to be sawn down and axed into oblivion and their roots pulled away by horses; alternatively, and more dangerously, they had to be pulverized with blasting powder or dynamite. Not a few workers lost limbs, faces, lives. But slowly the acres became clear, and with ploughing and tilling and raking, so the soil was revealed and prepared for the sowing and growing of crops. Aramaki took both to the local weather—cooler than in southern Japan, but nicely damp—and to the working conditions, with such enthusiasm and energy that by 1910 he had saved sufficient money that he was able to choose Towa Tasaki, a "picture bride" from a book of eligible ladies waiting back home in Japan, and to summon her to meet him in Bellevue. The pair married and three years later they had a child—the aforesaid Akira.

By now the Aramakis found themselves wedded also to the United States and to their new life in the Pacific northwest. Many of the thirty-odd thousand Japanese farm laborers who had sailed to the western states in those turn-of-the century years, when U.S. immigration policies were relatively liberal, wanted no more than to make money and return home, earned funds in hand. But a significant number bought instead into the idea of the American dream, and if they were farmers, or wanted to be, then to complete the picture, as the Aramakis did, they needed to buy a piece of land.

On the face of it, however, this was not possible. They were not citizens—they were Issei, a Japanese-sounding colloquialism but in fact an official American designation, signifying that they were first-generation immigrants, and as such could stay and work, but could not get U.S. citizenship. And if they were not citizens, it followed that they could not own land either: at the time, alien land laws were being passed in states all across the union, Washington State included, that firmly forbade Asians from acquiring real property—from becoming legal owners of any morsel of U.S. territory, of any size, location, or quality. The Aramakis certainly had the money to buy—their eight years of scrupulous saving had brought them the wherewithal to purchase a ten-acre piece just outside Bellevue. But the land office staff raised their hands helplessly, no matter how often Aramaki presented his billfold and counted out the purchase price: it was not possible, they said, being strictly against the law.

Except that there was, for the Aramakis and for many other equally frugal Issei in the region, a way around the rules. There was a loophole.

In 1921, the year that the Aramakis had rustled up enough cash to buy the plot, their son, Akira, was eight years old. Under the official designation he was Nisei, a second-generation Japanese, a child of Issei immigrants. But he was also something else. By virtue of having been born in the United States, and under the citizenship clause of the Fourteenth Amendment to the Constitution, he was an American citizen. He was what is called a birthright citizen—and by being so, he now had a greater claim upon the country of his birth than did his parents, since they were both born in Japan. And so it transpired that even if they, as Japanese immigrants, could not buy land in Bellevue, Akira most certainly could. He might not to be able to sign his name to a deed nor swear an oath before a notary public, but he could become the owner of a tract of land, notionally and, indeed, actually.

Which is how young Akira Aramaki came to be the legal and indisputable possessor of this ten-acre plot of stump-free and otherwise cleared farmland in an area of Bellevue called Midlakes. Once the paperwork was complete and the money handed over, then his parents, Hikotara and Towa, set themselves to work. Having cleared the land, they now began to cultivate it, initially growing small crops of vegetables, tomatoes, and vines, to bring in some income and to get the soil ready for the more complicated business of growing strawberries.

It took a couple of summer seasons before the fruit growing started to take off—but by 1923 it was doing so, and the Aramakis began to enjoy a certain very modest prosperity and some degree of standing in the community. Like so many of their neighbors and friends—and by some family estimates there were ten other Japanese families with property close by, such that almost all of Midlakes could fairly be said to be Japanese-owned— the family would either truck the glistening red berries across the lake to the markets in Seattle, have them sent down south to premium-paying customers in Portland or even San Francisco (although California soon imposed a strict quarantine on imported fruit and vegetables), or else sell them more cheaply in the local markets to the increasing number of summer visitors.

Life for the Aramaki family in those early years may well have been hard, but it nonetheless now held promise—not least because of the basic fact that the family now owned a tiny piece of the American countryside, a matter of inestimable meaning to any im-

Densho Digital Archive, 2008

The success of Japanese farmers in Bellevue led to an annual festival—with the election of a Strawberry Queen and her princesses—five young aspirants here—and the staging of classically American street parades.

migrant. And tough though their lives were, Hikotara Aramaki proved to be an industrious and inventive farmer, like so many of his neighbors, whose produce sold exceptionally well, being of good and reliable quality and—important for strawberries—sweetness.

Japanese farmers in America had form. Japan at the time was still an overwhelmingly agricultural nation, with smallholdings clustered in the fertile plains and a population well versed in the practices of small-scale farming. Not for the Japanese the business of managing waving prairie fields of golden grain; back home it was small rice paddy farming that dominated, as well as the growing of root vegetables for pickling, or the green-leafed legumes that were both so central to the local diet, and so restorative to the local soil.* All these growing skills they brought to the American west, and by doing so helped transform the local agricultural

---

* Daniel Green, the surgeon on Admiral Perry's 1853 expedition to Japan, was intrigued by the "peculiar hairy-podded bean, growing upon a branching stem, called commonly Japan pea" and he brought specimens back with him in 1854. American farmers had for sixty years previously been similarly fascinated by the plant's properties—especially its tolerance to cold and heat and indifferent soil—and began to cultivate these beans as a crop. They were, of course, soybeans, vastly important now to the American farming community, as they had been to Japan since antiquity.

scene while propelling the western states into the van-guard of the nation's food producers.

However, the land as the Japanese were legally able to buy or lease—just like the stump-filled barrens of Bellevue—was generally marginal, much of it floored with poor-quality soils generally disdained by the ex-isting farmers. The newcomers encountered all manner of natural adversity—blight, floods, locust invasions, near starvation—but in a singular fashion managed to come through them all, to the astonishment, admira-tion, and envy of many of the white farmers nearby. Yet envy can and did have poisonous consequences. A sig-nificant number of these white farmers fell prey to the burgeoning phenomenon of anti-Asian racism. Bodies like the Asiatic Exclusion League and the Native Sons of the Golden West, together with lawmakers from both main political parties, tried to put a crimp on the agricultural efforts of the immigrant Japanese.

And yet the Japanese managed to come through it all, hardworking, determined, united, and gener-ally optimistic. The statistics say much: in 1920 there were 3.5 million people living in California, of whom some 71,000 were ethnically Japanese, 2 percent of the population. They managed to produce fully 12 per-cent of California's total farm products, crops valued at some $67 million. By 1941, on the eve of war, their share of the truck farming crop of the western states had risen to an astounding 42 percent—even though they were still at no more than 3 percent of the population. Japanese farmers, working on holdings that seldom exceeded forty acres, produced 90 percent of the American west's snap beans, celery, peppers, and strawberries; and 25 to 50 percent of the area's asparagus, cabbage, cantaloupes, carrots, lettuce, onions, and watermelons. They had 30,000 acres laid to grapes, as well as 19,000 acres of plums, peaches, apricots, cherries, and almonds; they raised chickens in great abundance; and to top it all, they had control of some 65 percent by value of the flower market business. And all this was being raised and grown on the least fertile land in the states—a reality that, as supporters of the Japanese farmers were quick to point out, kept the prices of all these goods down, for the ultimate benefit of the western-states consumer.

But supporters of Japanese in those interwar days were vanishingly few in number, and the the drumbeat of opposition was growing steadily. Rival white farmers were now facing competition like never before, and bodies that represented them—the Farm Bureau Federation, as well as a national fraternal organization called the Grange—did their level best to head off the Japanese challenge. The Los Angeles County Farm Bureau lobbied the government to retract Japanese rights to lease, rent, or own not just farmland but "any land whatsoever." Valentine McClatchy, a wealthy and influential Sacramento newspaper proprietor, a prominent local landowner, and member of the more specifically named Japanese Exclusion League, was especially virulent. In 1920 he filed a court brief arguing that

the Japanese possess superior advantages in economic competition, partly because of racial characteristics, thrift, industry, low standards of living, willingness to work long hours with expensive pleasures, the women working as men etc. Combine with these characteristics extraordinary cooperation and solidarity, and the assistance of the Japanese Government, through associations acting for it or in its behalf, and the Japanese, concentrating in communities or industries, are easily able to supplant the white.

Such sentiments were not entirely new, and McClatchy himself had been voicing them since before the First World War. By that conflict's end, passions were starting to rage,* such that by the 1930s, with the job-destroying Depression now affecting so many sectors of American life—but not, notably, the Japanese farming communities on the West Coast—suspicion and hostility directed toward these people (who were now being called by some "agents of the proud Yamato race") became the leitmotif of the age. The new American president, Franklin Roosevelt, harnessed the FBI to look carefully at the Japanese population—by now swollen to a combined 120,000 in the four far western states of Washington, Oregon, California, and Arizona—and charged officials to draw up files and keep tabs. Transpacific geopolitical hostility was in any case starting to grow; the possibility was arising, as McClatchy and his like began to claim, that the Japanese farmers were no more than "a peaceful invasion force" financed and directed from Tokyo and bent on the single aim of coloniz-

---

*   An indication of the level of American official fear of a growing Japanese "threat" was the U.S. army's decision of 1919 to mount a transcontinental expedition by road to see how long it might take to move a large army to the West Coast to meet the challenge of an invasion by a then unidentified "Asiatic enemy." That supposed foe was Japan, identified by the National War College as a possible threat ever since its unanticipated defeat of Russia in 1905. Suspicion of the role of Japanese farmers in California—who were entirely innocent, as it later turned out—needs perhaps to be viewed in such a wider context.

ing the American west. They had to be watched, their every move investigated, their land grabbing—or other, more sinister—ambitions comprehensively thwarted.

Hikotara Aramaki fell ill and died in 1936— mercifully missing most of this. His son, Akira, by then twenty-three years old, continued to work the strawberry fields that he owned, along with his mother, Towa, who, while forty-six, was still fit and able. Life progressed more or less normally. Bellevue by now had developed into a sizable town, with many Japanese-owned businesses—grocery stores, furniture emporiums, druggists, barbershops. Mother and son would have been aware of the growing tensions between the United States and the Japanese empire, aware too of the vituperative and hostile sentiments of many of their white neighbors, but probably quite unaware that the FBI was watching them and their fellow Japanese. In October 1941, they would certainly not know that a special section of the FBI was drawing up a formal Custodial Detention List, categorizing Japanese into three levels depending on their perceived risk to the security of the United States. The Aramakis, quietly getting on with their lives, would be quite unaware that they were fast standing into danger.

And then came Sunday, December 7, 1941, and the Japanese navy's attack on Pearl Harbor. For the

120,000 men, women, and children of Japanese ethnicity who lived on the West Coast, everything suddenly and dramatically changed.

Those on the Custodial Detention List—most of them leaders of local organizations who were, somewhat fancifully, deemed a likely threat—were arrested on the spot by FBI agents. For the rest, like the Aramakis, mother and son, there were two months of agitation and nervousness, before President Roosevelt, on the advice of his army chiefs, decided to act. On February 19, 1942, he signed the infamous Executive Order 9066. This promptly sealed their fate.

The order declared the western halves of the states of Washington, Oregon and California, together with a smaller portion of Arizona, to be henceforward combined into one large military district, from which anyone could, on the order of the armed forces, be excluded. The order made no mention of any specific peoples, but its intent was crystal clear, and the intended sanctions soon began. All ethnically Japanese men, women, and children, whether American citizens or not, were ordered during that strange and frightening springtime to surrender themselves into army custody, to be sent first to a number of hastily created and temporary assembly centers. They would remain there, in generally execrable circumstances, while ten isolated concentration camps, most located in remote quarters of the west, were readied for them. These 120,000 people, together with such few items as they could carry, were then dispatched into the hot, dry, and dusty innermost wildernesses of America to spend the rest of the war locked up and rendered harmless, secure under military rule, no longer a danger to anyone. As though they ever had been.

Akira Aramaki and his mother, Towa dutifully did as they were told. Their assigned assembly center, which they reached by train late in May 1942, was named Pinedale, down in the blistering heat of the southern California desert close to Fresno. They were housed there for several tedious months—the tedium broken for Akira, as it happens, by the presence in the camp of a twenty-two-year-old woman named Hanako Tanako, originally from Sacramento, who caught his eye. The pair began a friendship which would lead to their marriage. But not quite yet: mother and son were suddenly dispatched without rhyme or reason back up north to a camp set down among the pillow-lava fields of far northern California, Tule Lake.*

---

* The concentration camps, immeasurably vast and cheaply built, were all sited west of the Mississippi River, well away from the prying eyes of the public and the press. There were two in

Here they remained for rather more than a year, before being transferred in September 1943 to the Minidoka camp in southeastern Idaho. Whereupon three things happened in quick succession. Akira and Hanako were promptly married by the camp commandant. Seven months later, on April 24, 1944, they had a child, whom they named Alan; and on the very same day their boy was born, Akira was released from camp and—despite being temporarily separated from his newborn son—began his long and protracted journey, ultimately getting back home to the strawberry fields of Bellevue.

Which is where he promptly discovered that his land was in the process of being taken away from him. And this was a circumstance which, for many of the returning detainees, was all too often the case.

Prewar hostility to the Japanese among the white

Arkansas, Jerome and Rohwer; two in Arizona, Gila River and Poston; two in California, Manzanar and Tule Lake; and one each in Utah, Wyoming, Colorado, and Idaho: Topaz, Heart Mountain, Amache, and Minidoka. respectively. Those of us who have visited all these wretched and forgotten places are unlikely to ever forget their names. But it remains one of the sorrier aspects of American education that the stark facts of the imprisonment of so many American citizens—for no reason other than that they were of Japanese origin—are barely taught in schools today, and that to most people in the United States, the names of the camps and what occurred there are almost wholly unknown.

The Minidoka concentration camp in southeastern Idaho housed some nine thousand ethnic Japanese during the war. It was one of ten camps established by the FDR administration as part of a policy since repudiated as a grave mistake.

farming communities of the west had now been well established, had gathered momentum more widely after Pearl Harbor, and had broadened still further during the detainees' prolonged absence in the camps—a period that coincided, of course, with the increasingly intense fighting against the Japanese in the Pacific theater, and the increasing number of white American casualties.

The hostility had been particularly unpleasant in the weeks and days in the spring of 1942 before the Issei and Nisei reported for their detention. The more callous of their white neighbors, eager for a bargain, bought the Japanese Americans' possessions—cars, domestic appliances, bicycles, furniture—at insultingly low prices, ransacked their shops, looked to move into their soon-to-be-abandoned homes, taking full advantage of their abruptly humiliating circumstances. To be sure, in many cases the white neighbors were kindly— promising to look after domestic pets, assuring the Japanese their houses would be looked after, their lawns watered and mowed, their possessions guarded, their lands preserved. The departing Japanese could do little else but trust the assurances they had been given, and hope for the best.

The situation of the detained farmers—of people like the Aramakis—was somewhat different from that of those who lived in the towns. For them, vital to the economy as producers of food, the government had strict rules. Up until the very day of your departure for imprisonment, you were ordered to maintain your farm and its lands as a matter of national security. You were bound by government instruction to keep on tilling, milking, sowing, ploughing, or combine-harvesting. If you failed to do so, either out of sheer ill feeling toward the government that was planning to incarcerate you or because you were too busy with getting ready to leave, you could be arrested and charged with sabotage, a seriously felonious breach of the law.

Akira Aramaki had not wanted to fall foul of this regulation, as he explained to an interviewer in the 1990s. As instructed, he applied the fertilizer, turned the soil, tied up the sweet peas, began to cut some lettuces. "I had my farm all ready to harvest," he said, remembering the early ripeness of his main crop of strawberries, in that warm early spring of 1942. "And then we had to go and sign up [for the camps] and everything. If I didn't get someone in there [to help], the government called us saboteurs, we were sabotaging.

"So I had to get somebody, and I got this good friend of mine, an Italian boy, [and] put him in there."

Aramaki might have seen the warning signs. He might have smelled a rat, once the tables were turned, once he was out of his property and the "Italian boy" was established there in his place. As he told the interviewer: "And gosh, the minute [he] took over—I wanted to stay in the house until I went into camp. And he even charged me rent. Yeah, and he's going to take over all my vegetables free of charge, and harvest the vegetables, and then he charged me rent!"

After which, silence. For the next two years Aramaki had no idea what was happening back at the ranch—until that April day in 1944 when finally the taxi from Seattle station brought him home.

The house, the fields, the outbuildings, all were in fine condition—except for the fact that the "Italian boy," now two years older and quite accustomed to his comfortable life in the Aramaki home, declared that he and his family were most certainly not going to move out. He was staying put. The farm was now his.

In his 1990 interview, Akira Aramaki reported that first encounter. "He said to me, 'The Japs aren't coming back; this land is never gonna be back—we're gonna have this land.'

"They took all my tractors, and all my irrigation pipe, and took it to their farm. They just ransacked the house. Because we had it all locked up in the rooms. And I didn't think they'd take it. But they took it."

Much the same was happening all across the American west in the mid-1940s. Camp by camp, person by person, those imprisoned were being set free, their detentions finally declared no longer a military necessity—not that they ever had been. Some, like Aramaki, were freed while the war was still raging; most were handed their money and their travel vouchers later, in 1945, when peace was finally declared. When at last they reached their various homes out west, their

relief was more often than not tinged with disappointment and disillusion. Anti-Japanese sentiment had now
reached a fever pitch in some communities; the houses
they had left behind had often been vandalized and
their possessions stolen; and in many a case the title
to the land a Japanese family had once possessed had
somehow vanished, like a will-o'-the-wisp, and they
found themselves just as landless as when their parents
had arrived, decades before.

The racial hostility may have been unworthy, and
the theft and vandalism unacceptably terrible. But the
misappropriation of land was on an entirely different
level, and quite unforgivable. To these migrants and
their descendants—as to all newcomers of whatever
race or persuasion everywhere—the acquisition of land
had an inexplicable, elemental importance. Those who
by hard work or good fortune had come actually to
own some acres were, simply by having done so, in-
eradicably bound to their new country. To have that
bond severed had a singular effect on those who came
home to find it so.

Of those Japanese, both older Issei and younger Nisei,
who lived in the western states immediately before the
outbreak of war, a little less than a quarter of them actually owned the acreage they worked. A very small number managed the farms for other, white, owners. The

great majority, some 70 percent, were tenants—and as such did not feel quite the same bond with the country, were quite literally not so invested in it. If, when they came out of camp, they found hostility and their house vandalized and their tenancy canceled, then this, despite its evidently traumatic nature, was a survivable change of fortune. Many of them could and did leave the land. They often quit the west. They started up again in many cases moving to a community or to a distant state where there was less evident hostility.

But this option was not quite so readily available for those Japanese who were, or who had been, actual landowners. Their emotional investment tended to act as a drag on their willingness to move. And so for a while, well on into the 1950s, many of these people, even if their land had been wrested from them, even though their life has collapsed around them, tried to stay on.

Their land had been taken in many ways, but no matter the manner, confiscated nonetheless. Most often in California—and most cruelly, since the victim of the confiscations were absent and held essentially incommunicado miles away in a concentration camp and quite unable to defend their own position—the alien land laws were brought into play. Legal arguments were offered by the state to the effect that the owners had no actual right ever to have owned the land, and so-called escheat proceedings were promptly initiated—just as if the landowner had died without a will and without any known beneficiary. If the lawsuits were successful—and they often were, since virtually no defense could be offered—then the land became the property of the state of California, which could dispose of it as it wished.

Other devices were employed. Tax regulations, in particular, required a lien to be issued if the landowner defaulted on his tax payments—and imprisoned Japanese often fell afoul of their creditors because the government had already frozen all their bank accounts, and they weren't able to pay their bills. An incarcerated Japanese might have the theoretical wherewithal to pay his taxes, but unless he could persuade a neighbor to pay for him, he was unable to do so, given that he himself had no access to his own money. The land then had to be foreclosed, on orders of the county or city to whom taxes should have been paid.

And then there was absolute theft (which seems to have been rare, the law providing some measure of protection) or else the much more fugitive notion of squatter's rights—whereby a person occupying the land, especially if he had done so for some while, could claim de facto ownership, or at least a right to remain on site, no matter what the real owner had to say about it.

Which last was the exact situation that Akira Aramaki encountered, when he arrived home that warm springtime evening in 1944. The Italian boy refused to move, point-blank. As it happens Aramaki did not crumple at his refusal, even though Bellevue was awash with anti-Japanese sloganeering, and even though his decision to stand up and fight to get back his modest acreage was initially greeted with jeers and derision. His battle took many months—but in the end judgment went in Akira Aramaki's favor, and he was indeed able to reclaim his house, and eventually to work his fields again, and to produce a modicum of crops, some of them strawberries, just as before the war.*

---

* Many others were less fortunate. For example, a lawyer named Takuji Yamashita had a spectacular run of ill fortune. He came to Washington State in 1892, proved a brilliant student at the University of Washington's new law school in 1902, but could not become a lawyer as he was a noncitizen, and the government refused to grant him citizenship. He and his wife then turned to farming. He found he could not own land for similar race-based reasons, and so persuaded a white neighbor to "own" the strawberry farm on his behalf. The Yamashitas were then sent off to a wartime detention camp and on their release found that, because they hadn't kept up payments, they had lost their farm entirely. He was compelled to work as a Seattle housekeeper for nine humiliating years before he and his wife returned to Japan in 1954, where he soon thereafter died, aged eighty-four. But there was a happier ending: in 2001 the University of Washington successfully petitioned the state to admit Takuji Yamashita to the bar, posthumously.

But by the summer of 1945, when his wife and child were released from the Minidoka camp, it seemed that his enthusiasm for agriculture was waning. Before long, and as with so many other Japanese farmers of his generation on the West Coast, he decided to abandon his chosen way of life, to sell up, change his career, and transmute himself instead into a classic icon of fifties American suburban prosperity. He became a real estate agent.

He died in 2004, aged ninety-one. His brief obituary in *The Seattle Times* recorded simply—and with a poignant elision—that "Akira was a lifelong resident of Bellevue, except for an interlude during WWII. He started as a truck farmer at Midlakes, retiring as a realtor." His wife, Hanako, lived on until 2016. Their son Alan lived out his working years as an engineer, remaining close to Bellevue and the eastern suburbs of Seattle. But his connection with agriculture was finished. That was all history.

Bellevue is a large and modern city now. The farms have nearly all vanished. The land of which Akira's father had been so proud, and which Akira himself had tended so carefully before he was taken away as a result of FDR's wartime policies, has long since been paved and cemented over. It now underlies a distribution center for the Safeway supermarket chain, and trucks roar in and out at all hours, bringing in and taking away boxes of cereal, detergent, lightbulbs, ice cream—and, no doubt, strawberries, grown on other tracts of well-watered land far away, and tended by people not Japanese, not anymore.

PART V

Annals of Restoration

1
Māori in Arcady

Last, loneliest, loveliest, exquisite, apart—
On us, on us the unswerving season smiles,
Who wonder 'mid our fern why men depart
To seek the Happy Isles!

—RUDYARD KIPLING,
the "Auckland" stanza of
"The Song of the Cities" (1893)

Change comes with great and impressive swiftness to New Zealand, and it does so for two reasons.

New Zealand was the last country on the planet to be discovered and settled by a human population.

The Polynesian canoeists were first, sighting the long chain of islands in, it is now generally assumed, the mid-fourteenth century. And then, three hundred years later, came the Europeans—with New Zealand's best-known ecologist, Geoff Park, noting drily that his country's fertile plains "were the last that Europeans found before the Earth's supply revealed itself as finite."

As well as being the last place on the planet to be found, New Zealand was also the first country on the planet to install as its chosen political system the most genuine kind of democracy, with voting rights early on extended to both sexes and to all residents—indigenous peoples included—quite regardless of their ethnicity. And further: Maori men were in fact given the vote in 1867, fully twelve years *before* their white counterparts, making New Zealand the first settler state in the world to give the vote to its indigenous population—and almost a century before Australia next door gave the vote to her aboriginals.

So, all of this country's human history, settlement to suffrage, has been compressed into a period of little more than five hundred years. Everything that happens and has happened in New Zealand's human evolution has taken place, when compared with elsewhere, at something approaching warp speed.

As recently as the 1970s, for example, there was little doubt about New Zealand's identity: everything about the place blared forth the notion that this was an indubitably and self-consciously British place, preserved in 1950s amber and set down comfortably in the southern seas. It was a place of daffodils and lawn bowls and afternoon tea, country constables on bicycles and cavalry twill and Marmite and well-attended Anglican churches. The nation's anthem was "God Save the Queen." And the name of the country was plain, simple, unadorned, giving just a nod to the Dutch sailors who first saw it: New Zealand.

But in the short span of the fifty years since, much has changed. The country's name now has an addendum: the Polynesian word *Aotearoa*, meaning "the land of the long white cloud," is now firmly co-opted into the old national title, and there are moves afoot to make it fully official. *Aotearoa New Zealand*: this triple-star of an appellation is what many like to call the country now. And the anthem—now officially and with the consent of Buckingham Palace coequal with the "Queen"—is also "Aotearoa," and the first of its five verses are an approximate Maori rendition of a locally beloved Victorian poem called "God Defend New Zealand."*

The anthem says much about the country and is widely held to be one of the most pleasing anywhere, remarkably free from jingoism and seen more as a prayer of humility than an exercise in nationalistic exultation, more a national hymn than a declaration of patriotism. The first time it was sung on a global stage was after New Zealand's rowing eight won the 1972 Olympic Games in Munich: the sight of the nine grown men in black standing and weeping as it was played remains etched indelibly into the popular New Zealand imagination.

Much more has changed than national titles and stirring songs, however. A very real effort was begun in the mid-1970s fully to incorporate into today's exceedingly white New Zealand—into today's Aotearoa, more properly—the 780,000 Maori people to whom

---

* Usually only two verses are sung in public, the first in Maori, the next the same first verse in its original English, and in that order. Since this convention was started in 1999, it is safe to say that all New Zealand schoolchildren now know the verse that begins with the words "*E Ihowā Atua, O ngā iwi mātou rā Āta whakarangona*"; most others also know the English "God of Nations at thy feet, In the bonds of love we meet."

the country should—and indeed, once did—belong. This change in emphasis, which percolates through all aspects of the country's society today, and while well intentioned is not without its failings and its critics, began with what is now a concerted national effort to answer and deal with this most fundamental question: *Who exactly owns the country's land?*

There are 66 million acres of New Zealand, from the tropical tip of Cape Reinga in the north to the cold and blustery southern cliffs of Stewart Island, a place of tussock grass and marsh so bleak you know it is just a short hop from there to the bitter gales of the Roaring Forties. These millions of acres, in the main lushly emerald pastureland and spectacular (but alarmingly seismic) mountain country, constitute a land that was seized, wholesale, by those British outsiders who first found it for themselves and then, after Captain Cook's first claim to the place in 1769, eventually came and settled there. Much of this land now needs, in the eyes of many in New Zealand, to be handed back.

To the extent that talk about such a possibility is occurring at all, and to the extent that the need for a restoration of Maori right is now part of the country's national will, the reforms that are under endless and definitely not warp-speed consideration surely bode well for the moral compass of the nation. Many locally will suggest that in terms of social justice New Zealand is an impressively successful place, on myriad levels. And to those who think so, the country's programs that deal with native land are seen as central to such success and reputation as the nation currently enjoys.

The key event in the history of New Zealand, most especially her complicated relationship with her 66 million acres, was the signing on February 6, 1840, of the Treaty of Waitangi. This remains the central icon of the country's national origins myth. And given what has happened since, it seems very much more important today than it ever was back in the nineteenth century.

The physical treaty itself still exists, as much revered as elsewhere are Magna Carta and the Declaration of Independence. Unlike these two very much more venerable documents, the treaty is now a decayed, ragged, and rather careworn thing. Nonetheless, it is sedulously preserved and guarded, its fragments on view in glass cases in a darkened room in the National Library in Wellington, a hallowed and hushed place.

The events leading up to the treaty's making and signing took place at the zenith of Britain's imperial adventuring—in the times when, as previously noted of the locals, "we opened our eyes, we had the Bible and they had the land." The trope that has long since defined British missionary behavior in Africa was much replicated in the Antipodes: James Cook—bent on a mission of a different kind, but a mission nonetheless—stumbled across island after island in the Pacific, and he took them, annexed them, claimed them (the language varies) for Great Britain. He came to New Zealand in 1769, then sailed on to what would be New South Wales in 1770, and notwithstanding the presence of Maori people in the former and aboriginal Australians around Botany Bay in the latter, he declared all now to be under the sovereignty and benign invigilation of Britain. He then sailed away, discovering and claiming as he went, to tell his faraway sponsors of their new austral possessions, and leaving the locals, to the extent they understood what had happened to them, mightily dazed and confused.

Migration to Australia—initially, and infamously, of convicted and impoverished petty criminals who were sentenced back in Britain to transportation to this new-formed penal colony—began in earnest in October 1788. New Zealand, however, was initially discounted as a destination—it was thought neither suitable for prisoners, nor tempting enough for voluntary pastoral-ists, and so it was retained in a loose association with New South Wales, administered informally, more or less forgotten about.

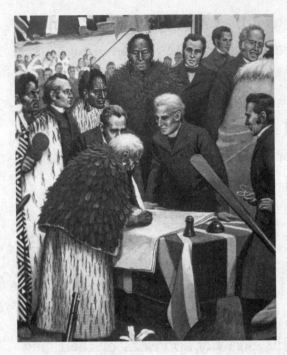

The Treaty of Waitangi, first signed in February 1840 between British officials and Maori chiefs, formally established British sovereignty over the island. Interpretations of its various translations have led to much controversy.

A handful of Europeans settled, equally informally. Some few were convicts who managed to escape from the relatively nearby British prisons in Tasmania and on Norfolk Island. And then once the Chinese trading houses in Canton said they would accept sealskins in exchange for tea,* a ragtag huddle of British seal-

---

* So popular was Chinese tea in London that the British East India Company, which had a monopoly in the trade, had run out of metal coin as payment, and in the late eighteenth century was starting to employ opium as a medium of exchange. This infuriated the Peking mandarinate, a development that led to the

ers set themselves up in the South Island, in the vain hope—augmented by some onshore whaling, once the great beasts had been spotted cruising outside the west coast fjords—of making great riches. Many of these sealer-settlers lived with Maori women—with no objections from the Maori menfolk, who were generally kindly and accepting. A scattering of missionaries then arrived and, after an initial sturdy reluctance, by the 1820s some Maori did agree to take baptism. And then not a few Maori—whose customs and practices the British found intriguing and admirable, most especially relevant here being their view that all land is owned communally, and not by any individuals—were invited across to Australia, to be properly "civilized," there to acquire a working familiarity with the English language, just in case Britons ever decided to come to New Zealand en masse. But they did not come, at first, and in the first few decades of the nineteenth century New Zealand was still almost wholly Maori and was generally omitted from the colonial gazetteers. As if to rub the point home, New Zealand was particularly mentioned in an 1817 law, the Murders Abroad Act,

---

Opium Wars, the cession to Britain of Hong Kong, and the eventual public humiliation of imperial China. One imagines that the sealskin trade would not have had such consequences.

which held that murders committed in places *not* then under British rule—the United States, Tahiti, Honduras, *and New Zealand* among them—would be treated as if carried out on the high seas.* It was not a colony in law—although by 1830 it started to be accepted by London that such a growing number of Britons were now beginning to live there, dotted around the country, that its status should be reconsidered from time to time.

Time to time happened soon thereafter. Matters started to change in 1833, when the British decided to act. They first, and somewhat incautiously, sent down an ill-suited emissary, a professional wine grower named James Busby, who was charged in his new-made post of British Resident with managing the island outpost on London's behalf. He landed in the Bay of Islands, and being generously welcomed by the local

---

* The law's reciprocal, the Colonial Prisoners (Removal) Act of 1884, allowed prisoners of the empire to be brought back to the United Kingdom. In 1985 a senior British judge traveled to the mid-Atlantic colony of St. Helena to help retrieve an inmate from the tiny Jamestown jail, which was considered too small to hold him. Some 170 years earlier Napoleon Bonaparte had rather larger quarters during his incarceration on the same island, after his Waterloo.

Maori chiefs set up his headquarters in the estuary village that would later become so famous, Waitangi.

His appointment turned out to be no great success. Rather than sticking to his formal remit, Busby apparently became enamored with the idea that the Maoris might create for themselves some form of properly organized government—ignoring the fact that since so many of the various Maori tribes were at odds with one another, the establishment of some kind of central government was unlikely. Nonetheless, this was the Busby plan, and he further supposed that, once self-governing, the country's leaders might form themselves into a wholly independent nation. His optimism was such that he even had their leaders choose a flag for themselves.

London, on hearing of these developments, and also hearing talk of violence breaking out among various gatherings of British settlers in South Island, dismissed their Resident's ideas as plain silliness and tomfoolery. Generously, though, they assumed publicly that by making such grandiloquent plans Busby was merely trying to see off the rival French, whose imperial tentacles were starting to curl down from Tahiti toward New Zealand's North Island.

As Busby's dispatches became ever more irritating, and as the trickle of inbound Britons continued to grow—two private settlement companies had been formed back in London for the express purpose of encouraging migration to New Zealand—London decided to act: all dithering would henceforth be ended, Britain would formally annex the islands once and for all, and create a new colony. A naval officer named William Hobson was given orders to head off south to Waitangi and formally to annex the islands for the newly enthroned Queen Victoria.

Hobson was already familiar with the region: he had served as commander of the HMS *Rattlesnake*, employed during the founding of the settlement of Melbourne; and he had been summoned across the Tasman Sea in 1837 by the vexed James Busby to put down a small Maori rebellion and a simultaneous outbreak of lawlessness among some of the settlers. Both had fizzled by the time Hobson reached the territory, and so instead he wrote a report on the situation to give to the Admiralty once he was back in London. But then he was told, essentially, that he needed to change his clothes and head back down south again, this time on imperial business. His sailing orders were initially simple, though to the confusion of many they changed while he was en route. In consequence, by the time he arrived in the village of Waitangi (whose name means, appositely, "the waters of lamentation") he had written a treaty document that would lead to a raft of problems—and, to be fair, a raft of possibilities as well—that are still very much in existence today, two centuries later.

His original instructions—written by a group of senior Colonial Office bureaucrats, evangelical Christians who had led earlier movements to abolish slavery in other British possessions, and who were broadly sympathetic to the Maori people—were straightforward enough. Hobson would apologize for disregarding the earlier Busby self-government plan, but would instead declare that Queen Victoria was now formally claiming sovereignty over New Zealand and her native people; and that Maori land would be purchased, at fair prices, by the colonial government. The land would then be resold to would-be British settlers, with the revenue from the sales being used to finance the colonial government's operations.

But as he got under way, the Colonial Office amended the arrangement. It had come under immediate pressure from the two private settler bodies that had been created to encourage migration, which had different views. Rather than Hobson heading south to create a system whereby British settlers would be accommodated in a country that was *still broadly Maori*, he should make instead *"a settler New Zealand"* in which *"a place had to be kept* for the Maori." And so the treaty he was to write—and Hobson was given no official help to compose what would in effect be the new country's foundation papers—should reflect this changed priority: settlers first, Maoris next in line.

Considering the profound eventual importance of the document, it remains remarkable that Hobson, together with the already in place James Busby, was able to hammer out the document in just four days, all in the comparative comfort of Busby's seaside cottage. It had then to be translated into Maori, a language with which few were familiar—the local specialist being Henry Williams, an Anglican cleric. Williams achieved that task overnight, but it is his extraordinarily rushed rendering that lies today at the heart of the many problems the treaty has since thrown up.

On the midsummer Wednesday morning of February 5, 1840, the local chiefs and hundreds of ordinary Maori were summoned by runners to attend a formal reading of the treaty. A giant marquee had been erected on Busby's front lawn, and standing on a hastily made dais Hobson read the English text aloud. He did so for posterity of course, but also for the benefit of the scores of settlers who had turned up, though they were told firmly to remain silent during the proceedings, since this was an event exclusively concerning relations between the Maori people and the British Crown. Some French had arrived as well, eager to torpedo any arrangement involving the British, trying to persuade the Maori to convert to Catholicism and extolling the virtues of being ruled from Paris rather than London. Few paid them heed.

Over a gathering din, the Reverend Williams then calmly read out his translation,* paragraph by paragraph, to the Maori crowd—men who, according to the other visitors, seemed initially quite unimpressed by what they were hearing, though whether it was the content or the grammar that most concerned them is not known.

The Treaty of Waitangi, after a formal preamble telling the chiefs of Victoria's pledge to protect the Maoris' rights and property, had just three articles—with which the chiefs were "invited to concur," by signing.

---

* What Maoris know as *te reo*, "the language." is not dissimilar to Tahitian and other East Polynesian tongues. There was no written version until a polymathic Cambridge linguist, Samuel Lee, devised a comprehensive fifteen-letter orthography in 1820. Almost three hundred Maori words are recorded in today's English language, most of them signifying varieties of plants, fishes, or birds, but including the pre-rugby-game terror-dance known as the *haka*, the amiable greeting *kia-ora*, and, of course, most emblematic of all, the shy and flightless *kiwi*.

The first required "the Chiefs of the Confederation of the United Tribes of New Zealand" to cede to the Queen of the United Kingdom "absolutely and without reservation all the rights and powers of sovereignty . . . over their respective Territories." A simple enough imperial demand as things go, making the Queen, Queen of New Zealand.

The third article, just as unambiguously simple, gave all New Zealand native peoples the full rights of privilege and protection as any other British subjects. That might in due course suffer some limitations and restrictions, but on the face of it, for now it seemed plain enough.

It was the second article, dealing with land, that would eventually cause all the trouble and consternation. It stated that "the Queen guarantees to the chiefs and tribes and their families the full exclusive and undisturbed possession of their Lands and Estates Forests, Fisheries and other properties . . . so long as it is their wish and desire to retain the same in their possession." The chiefs would also give exclusive right to the sale of their land to the Crown or the Crown's official representatives.

One of the many failings of the British empire, when making such attempts as this to bring a native people gently to heel—rather than the more traditional means of colonial expansion via conquest and annexation by force of arms—was the often bovine inability of Britons to even try to understand the subtleties and nuances of another and unfamiliar people. The Maori chiefs assembled here, studying the translated text that was presented to them, could have had no notion whatsoever of sovereignty, nor could they fathom what the rights of British subjects might be, since they had no certain idea of what or where Britain was. And as for phrases like "the full exclusive and undisturbed possession" of lands—to a Maori the concept of individual land possession simply did not compute.

One Maori word used in the treaty continues to cause trouble. The British first granted to the chiefs what in the translated text was called *rangatiratanga*—from the word *rangatira*, meaning "chief"—and which was taken by all to signify "chieftainship." But the colonists also went on to ask the chiefs to cede to their new rulers *kawanatanga*—and this was an invented word, a neologism, and it was intended by the British intended to mean "governance." The chiefs agreed to both terms—a concurrence which historians suggest means that the Maori kept something that meant a great deal to them, but at the same time ceded to the newcomers something that meant very little, or of which they had no real understanding.

The kinds of questions the assembled chiefs put to the white men displayed their perplexity: "What do we want of a governor?" asked a leader named Rewa. "We are not whites nor foreigners. We are the governor. We are the chiefs of this land. Return!" The very concept of empire that was embodied in Hobson's visit—let alone its implications for the solid surface on which all the locals lived, on the forests and the fields and the seashore and the mountain ranges—eluded all the Maori that day. The settlers got it, of course, as did the French, whose Gallic disdain for the British was all too evident. But the Maori, presented with this document, must have wondered why on earth they were being so bothered, why shaken so rudely out of their long arcadian slumber.

Nonetheless, after a day and night during which the various Maori groups argued among themselves—with Henry Williams doing his level best as intermediary— the colonists' coercion, so silkily applied, finally achieved its goal. On February 6, the assembled chiefs informed Hobson's men they were ready to sign. Hobson himself had retired to the ship that had transported him from Britain, HMS *Herald*, and had not expected an agreement so quickly. Stories still circulate suggesting that he arrived at Busby's house in a state of undress, and signed the British side of the agreement while in his dressing gown. (Paintings of the event naturally show him in full naval uniform.)

The first Maori to sign was one of Henry Williams's converts, a Bay of Islands chieftain named Hone Heke Pokai, who was able to sign his name. Twenty-five others followed him—such that by the end of a day observed as a holiday every February 6 in today's New Zealand—it could fairly be said that over half of the senior chiefs who were present that day had agreed, with their signatures or their marks, to London's imperial demands.

Even so, Hobson thought it prudent to take the proposed treaty around the country to win the approval of other Maori leaders. So he had eight further copies run up, each of them handwritten and covered with official stamps and seals, and all translated with more or less the same text. By the time this exercise was complete, more than five hundred chiefs had added their various notations of concurrence, and by the time the austral spring was bringing the daffodils back to Otago—one of several South Island regions that never got to participate, since the weather was too foul for the treaty carriers to visit—the deal was done. William Hobson then formally detached the country from the superintendency of New South Wales, declared himself governor of the Crown Colony of New Zealand, and swore in three fellow Britons to perform the territory's senior management. With a budget of four thousand pounds, a bureaucracy of thirty-nine junior civil servants, and a gathering of eleven policemen* brought over from Sydney, the new governor proceeded to run the colony as best he could.

One of Hobson's first appointments was of a Land Claims Commissioner, an English lawyer named William Spain. However important matters of sovereignty and protection might seem in principle, the most significant practical business of the Treaty of Waitangi concerned the disposition of the country's 66 million acres of hitherto communally owned land. That would produce the very raft of problems that are being dealt with in New Zealand to this day—and which other nations with substantial indigenous populations are watching closely.

First of all, though, argument over the interpretations of the treaty triggered a series of vicious land wars between Britons and Maori, conflicts that would last for thirty wretched years. These confrontations became so serious that they would eventually involve the summoning over from Australia of eighteen thousand Brit-

---

* Reportedly alcoholics who might otherwise have been retired from the service.

ish soldiers, complete with artillery pieces and cavalry regiments. Those troops would put down rebellions of a ferocity quite unimagined by those who had so congenially signed the paperwork back beside the seaside in Waitangi. The passionate intensity that underpinned the Maori attitude toward their land—it could not be for sale because it was not owned by anyone—was all part of the subtleties of another culture that seemingly so often eluded the Britons' understanding. Nearly three thousand people, by far the majority of them Maoris, died as a result of the fighting.

Numberless battles of various shapes and sizes were fought during the mid-nineteenth century and pitted rebel Maori groups against the British colonial government, although matters were often complicated by the presence of loyalist Maori fighting alongside the British and so against their own people. The rebel group, known in the vernacular as the Kingitanga movement, sported a homegrown monarch who presided over all tribes and who was intended to put the locals on an equal footing to Queen Victoria. The British naturally wanted to squash such a potential threat to their governing authority and brought in those Australian troops to put it down. The biggest battles were fought south of today's Auckland, in a region called Waikato, where the British had to send in a naval vessel to help dislodge some of those they regarded as "enemy" forces. A proclamation was first issued by the colonial governor, its closing lines reminding the Kingitanga supporters what this fight was really all about:

> Those who wage war against Her Majesty, or remain in arms, threatening the lives of Her peaceable subjects, must take the consequences of their acts, and they must understand that they will forfeit the right to the possession of their lands guaranteed to them by the Treaty of Waitangi.

The British won the war, eventually, but suffered such bloodied noses that they promptly began a legally mandated program of land confiscation, seizing a total of 3.5 million acres of Maori land purely as revenge, as small-minded punishment for the natives ever having the gall to challenge the British. The confiscations left a legacy of hostility between Maori and *pakeha*—white men—that festers to this day.

The situation in the New Zealand at the beginning of the twentieth century was thus deeply unsatisfactory and dominated by these two land-related issues: a carelessly written, poorly and ambiguously translated, and ill-thought-out treaty as the nation's founding instrument of existence; and a program of land confiscations that seemed capricious, vengeful, and pointless. In the early days the Maori had been in a majority; now they were a minority in their own country, easier to ignore and to bypass, and an increasing number lived lives of greater and greater poverty as the years went on.

Various laws enacted during the beginning of the century had already taken large tracts of their land away—for national parks, for defense establishments, for reasons of conservation. And in the aftermath of the Second World War the country witnessed a series of episodes of very obvious racism: many Maori farmers found themselves tragically unable to get development grants—while white, *pakeha* farmers experienced no such problems. This all too often forced many Maori to sell their farms and fields, and move into the cities. This wholesale shift in population had dire social effects, but it also mightily enhanced the white ownership of the country, and steadily reduced the portion that was owned by Maori tribes—a people who, before 1840, had owned everything.

They did win some small victories, to be sure: in the 1920s a land commission ruled that some of the confiscations and fraudulent purchases were indeed unlawful and offered compensation. However, the sums offered were generally niggardly and were rejected as insulting. Meanwhile, the mean-spirited actions of the *pakeha* rulers went on. In 1953 the government declared that any unused Maori land could be leased, long term, to white people. Fourteen years later, in 1967, an even more extreme piece of legislation allowed any Maori-owned land that had four or fewer names on the deed to be compulsorily reclassified as general land, up for grabs by all.

All the while, the notion that the Maoris' ever more straitened circumstances had a great deal to do with the steady purloining of their lands was simmering beneath the surface. In the middle of the 1970s it could be contained no longer, and pressure for change started to build, the Maoris seeing their situation as verging on the intolerable.

At which point there entered into the story an elderly Maori widow, a lady of rare eloquence and determination named Whina Cooper, who managed by sheer force of will to galvanize the dejected Maori community and turn the situation smartly around. A flinty, acerbic, and often divisive presence even within her own community, Whina Cooper—her first name was pronounced "Fina," and came from Josephina, given at her Catholic baptism—is still seen by many as the nation's mother figure. During her long life—she died in 1994, just shy of making her century—she was variously a teacher, a rugby coach, a champion of breast-feeding, a demon shot with a rifle, a leader of her tribe, and a kauri-gum digger.*

But her passion from childhood was the need, instilled in her by her militant father, for Maoris to win back their land. Her campaigning life began while she was still a youngster, when she sabotaged a white farmer's drainage pipes after he had stolen estuarine mudflats near her childhood home and tried to develop them for growing crops. Her final and best-known campaign, undertaken in the austral spring of 1975, when she was eighty years old and frail-looking, had her leading a march of Maoris and sympathizers through the entirety of the North Island, seven hun-

---

* The gigantic kauri trees found in the far north of New Zealand ooze a particularly viscous resin which tends to collect into pods or lenticules eventually to be buried many feet down in the earth. Since the gum was much prized for making varnish and in the production of linoleum, a sizable nineteenth-century industry grew up, whereby gum diggers, using eight-foot-long hooked spears, prodded the thick soil for pods of gum, which they would then extract, wash, and send down to Auckland for export. The industry particularly attracted a large number of migrants from Europe's Dalmatian coast, and to this day being a descendant of Dalmatian gum diggers is a badge of honor in New Zealand's North Island.

The formidable octogenarian Whina Cooper led a month-long march of Maori activists down through the North Island to parliament in 1975, a protest that led to major changes in the national land policy.

dred miles of marching down to the capital to protest at the steady and government-sponsored dissolution, as they saw it, of their territorial birthright.

Just fifty activists set off on the Maori Land March on a mid-September Sunday, with Whina Cooper, in her trademark headscarf and equipped with various items of traditional ceremonial regalia, leading the group. "A Landless Maori Is a Lost Maori" read one of the posters, summing up the sentiment. Each night the marchers stopped at a *marae*, the gathering place of a local tribe, where they were fed, watered, and rested. Each subsequent day more marchers joined, and after twenty-nine days of steady progress, often in poor spring weather, some tens of thousands reached their intended destination of the Parliament building in Wellington. There the weary and wet Whina—it was raining heavily as the group surged through the notoriously weather-bound capital—presented a petition signed by sixty thousand Maori, demanding once and for all a comprehensive reform of the land laws. "No more sales to *pakeha!*" some cried. "Give us back our land!"

Without a doubt the Land March was a tipping point for the status and self-regard of New Zealand's Maoris. They had been treated in the most cavalier fashion, pushed to one side by the ceaseless tides of white immigration, and they had also suffered the relentless loss of a large proportion of their spiritually precious land. Now, something had changed. The best-remembered image from the march, of this tiny, nut-brown, weather-beaten figure of a venerable old lady tottering gamely along an empty, rainswept stretch of roadway, with herself so possessed of a fanatical passion for a long denied fair deal for her people, somehow lit a fire. Sympathy for the Maoris' plight became a national trope almost overnight. In cultural terms, the march was judged a considerable success—of that there was no doubt. It is notable, for instance, that it was around this time that citizens of all stripes began to refer to their country more widely as Aotearoa, and the anthem to be sung as an alternative to the "Queen" had its first verse now sung in Maori, a tongue with which all children of the seventies, *pakeha* among them, would soon become passingly familiar. Maori names, artifacts, and customs started to become an integral part of New Zealand society, with the country's common language soon incorporating more and more scatterings of Polynesian words and phrases.

But as for the principal reason for the march—the need for the government seriously to address Maori grievances relating to their land—matters are extremely complex. Nevertheless, progress has most definitely occurred. The reforms enacted in New Zealand over the years since the Land March have been considerably more substantial than in those other settler states—in particular, the United States, Canada, and Australia—where an indigenous population has been well-nigh obliterated by an immigrant horde. There have been some small but signal successes for the New Zealand land rights movement, both atmospheric and actual.

For example: as the march was reaching its halfway point, on October 10, the government announced the establishment of a tribunal that would formally investigate alleged breaches of—or, because its shoddy translation, any disputed consequences of—the 1840 Treaty of Waitangi. Many Maori leaders, who after suffering through years of government perfidy had ample reason to be skeptical of the idea, were fearful that this Waitangi Tribunal would be just another device for allowing official delay or disdain. But as it happened the sea change in attitude that New Zealand was already undergoing in the mid-1970s resulted in the tribunal becoming an instrument of some real and lasting value.

The act setting up the tribunal has been amended no fewer than five times since 1975—an illustration of the speed with which things can change in New Zealand if minds are put to it—in attempts to widen the tribunal's scope and to answer various complaints about its limitations. Most notable of all these was the amendment made in 1985 that allowed claims to be investigated as far back as 1840, to the date that the original treaty was promulgated and signed back in James Busby's seaside cottage in Waitangi. That single change proved to be a decisive, almost revolutionary step—for it meant that the entire meaning of the founding document of the colonized nation could now be challenged by those who had been colonized. In law at least, the Maoris for the first time had some semblance of possessing the upper hand.

Claims promptly rocketed into the stratosphere. During the first ten years of its existence, the tribunal saw no more than half a dozen claims annually.

Once the 1985 amendment came into force, allowing the original treaty and all subsequent events to be challenged, at least a thousand claims were immediately filed—and a substantial number of them were accepted and adjudicated in the Maoris' favor.

One of the more spectacular claims was made for the yearlong outbreak of warfare that raged in Waikato in 1863, which was followed by a brutal series of confiscations made by the colonial government, as punishment. The tribunal agreed that the confiscation had been wrong, awarded the local Maori tribe involved some $170 million, and, most significantly of all in terms of its symbolism, had Queen Elizabeth issue—in person, during a 1995 royal tour of New Zealand—a formal written apology. Seldom does a sovereign ever say sorry;* the Maoris took her doing so in 1995 seriously, and with much solemnity and gratitude.

---

* The local Tainui tribe had asked that the queen speak her apology, and that she do so on Maori-owned land in Waikato, but Buckingham Palace said no. Her decision instead to sign, in triplicate, the act authorizing the compensation and the apology was said to be the first time in her reign that she had ever signed a law in public. The fact that while doing so she wore her traditional Maori *korowai*, woven from thousands of feathers, was seen as giving appropriate respect—*mana*, in Maori—to the occasion.

During her visit to New Zealand in 1995, Britain's Queen Elizabeth II, wearing a traditional Maori feather cape, signed an unprecedented formal note of apology and compensation for Victorian-era land-grabs.

There have been scores of successful claims since—the best-known being the so-called Sealords settlement of 1992, which paid some $100 million in compensation for the previously denied Maori ownership of the country's fishing rights—and gave Maoris a half share of a major new fishing company. In 2008 a further $150 million was awarded for the country having denied locals revenue from forests that stood on lands that were once Maori. All told, some $600 million has been paid out in various kinds of compensation to various Maori clans; in addition, a number of famous New Zealand place-names have been changed—most famously, Mount Cook became Aoraki, "Cloud Piercer." Maps of the new New Zealand's 66 million acres are now liberally decorated with changed place-names; and the words themselves have been ornamented with more macrons than can ever have been imagined. Maori, for example, is now unashamedly Māori, on all official documents.

New Zealand is now a changed place indeed. The land is still not fully shared—and a newly rejuvenated Maori political party said in 2020 that it would campaign to have the ownership of *all* New Zealand's land put up for negotiation, which would be mightily complex an affair, likely to be bogged down in courtrooms for scores of years to come. And only if, that is, the Maori party won a majority in Parliament—which seems improbable at best. The party's proposal is thus probably a nonstarter, and its members will most likely have to come to terms with the reality that if the land is still not as apportioned as right suggests it should be, it is certainly more equitably shared than when the settler-missionaries first came, when they gave the local people the Bible and told them to close their eyes and pray. Such change as has taken place has been for the better, which is what to most in today's New Zealand it seems to be. It has also come swiftly, as perhaps befits a place with so very brief a human history, and one that is so far away that it's been able to do its own thing, without interference from neighbors.

2

Strangers in the Hebrides

On those islands
Where no train runs on rails and the tyrant time
Has no clock towers to signal people to doom
With semaphore ultimatums tick by tick,
There is still peace though not for me and not
Perhaps for long—still peace on the bevel hills
For those who still can live as their fathers lived
On those islands.

—LOUIS MACNEICE,
"The Hebrides" (1937)

The Scottish island of Ulva is small, is almost wholly unpeopled, and, if not exactly as distant as New Zealand, is still to most people in Britain very far away.

Depending upon whether you approach Ulva from the south or the north, getting there requires passage on either two or three ferries, since the island lies a few hundred yards off the coast of the island of Mull, and there are no bridges to either to Mull or to Ulva, nor probably ever will be.

Ulva and its 4500 acres of farmland used to belong to one person. It also used to sport a population of six hundred people—at peak times, eight hundred. Now it has a population of just six.

But matters have lately changed, profoundly. And, significantly, they have changed in large part because of government fiat. Because of politics. Since the summer of 2018 this island that used to belong to one person has come to belong to many. By undergoing this change of status, Ulva has become a poster child for a new form of ownership that has swept through Scotland in recent years, one that is summed up by the rallying cry of the landowning revolution it represents: it has suffered, or has enjoyed, a *community buyout*.

The most recent owner was Jamie Howard, who in Scottish terms was known as the Laird of Ulva.* But as

---

* A laird is generally the owner of an estate and has no significance beyond the place that is owned. It is not a hereditary

it happens his own island life was blighted by tragedy—maladies and deaths of those close to him—such that he came in time to consider Ulva somehow accursed, and the lairdship that he held fit for abandonment. Inconsolably dejected, he decided to move away for good, and to put the island, along with the big family house, a cluster of uninhabited farm buildings, and various well-worn stone cottages, some of them housing his scattering of tenants, up for sale.

He had to wait only a short while before offers started to tumble in. Islands in Scotland seldom come onto the market; Ulva, a place of a singular sea-washed beauty and steeped in Gaelic history, promptly became a fragment of the rarest kind of real estate, a thing of great desire—such that the first prospective buyers came in by helicopter, so eager were they to realize a quick sale.

Jamie Howard was soon within an ace of concluding a deal for several millions pounds, with which he could begin a new life well away from the melancholy memories of the Inner Hebrides. He was impressed with the man who wanted to buy, regarding him as a potentially fine steward for an island that he and his

---

peerage, like a duke or a viscount, and is merely conferred upon purchase and transferred to a new owner upon sale.

family had long regarded—despite its recent mournful history—as a unique place, deserving of being long cherished.

However, matters did not turn out as Jamie Howard and his buyer had hoped and expected.

Scotland had been grappling for decades with what is seen by many of its five million inhabitants as the grossly unequal ownership of its land. In the poorer quarters of the great Scottish cities that W. H. Auden referred to in his poem "Night Mail" as "working Glasgow . . . well-set Edinburgh . . . granite Aberdeen," there was smoldering resentment that so much of Scotland's cherished landscape, in the Highlands and islands particularly, belonged to so few. A polemical drama group known as the 7:84 Theatre Company was formed in 1971, its name derived from the then widely accepted figure that 84 percent of Scotland's land was owned by just 7 percent of the people. In more recent times, those statistics have hardly changed. According to one study just 432 families are said to own half of Scotland's present landscape; according to another, 1200 families own two-thirds of the national surface. Over the years there were regular bouts of fulmination and bluster coming from left-wing commentators and members of Parliament—but relatively little real legislative notice was ever taken, in large part because the Parliament of the time (and indeed, ever since the 1707 Act of Union) was sited down in London, many hours away from the supposed problem, and in England, quite another country.

All this changed, however, in 1997, when a referendum staged in Scotland resulted in the establishment for the first time (for three centuries) of a separate devolved Scottish Parliament, one that would be established in the Scottish capital city of Edinburgh. By the turn of the century, the institution was fully established, a unicameral body with 129 seats, meeting initially in a church hall and, since 2004, in a suitably grand new building.

From its inception, left-wing parties dominated the newly established body, and in 2003 its members passed legislation that encouraged the communal ownership of land—a move that had the effect of starting the dilution of some private holdings and establishing, especially on some of the larger islands, a more dispersed system of ownership. But in the view of the more radically minded, this law dissembled, did not go far enough—it encouraged but it did not mandate communal ownership. And so matters did not start to change profoundly until 2011, when the Scottish National Party (SNP) broke through the nativist glass ceiling, won a majority in Parliament, and began to bulldoze its reforming way through the country's social system.

Of all the SNP's various revolutionary plans for Scotland, the one that in the view of the party chiefs would most obviously allow the country to become more equitable was always the plan for fundamental land reform. Not the near-beer reform the traditional political parties had espoused—but full-blown radical reform, which would really shake things up. Now, with the SNP firmly in charge and with a comfortable majority in Parliament, a plan for real change could get under way. "Scotland's land is a valuable asset," the newly elected SNP member for Argyll remarked, "and an asset which should benefit the many. The party is committed to empowering communities through land reform, enabling them to determine their own futures."

A million acres would be in community hands by 2020, pledged one minister. All land would be fully registered within a decade, promised another. Quiverfuls of extreme reformist legislation would now be let fly, to make sure the land in Scotland was henceforth to be distributed among the people more fairly than ever before.

The result of all this pledging and promising was the swift passage of the Land Reform (Scotland) Act of 2016—a draconian piece of legislation that, for some landowners, had an immediate, chilling effect.

The earlier act of 2003 merely encouraged communities to bid for any landed estates that came up for sale. This 2016 act went further: if the would-be new owners, communities all, pledged to indulge in the loosely defined "sustainable development" of their purchase, then the law positively forbade the current landowners from selling to individuals, but compelled them (if they planned to sell at all, that is) to sell only to such forward-thinking communities. It was the passage of this powerful new law allowed the SNP's charismatic new leader, Nicola Sturgeon, to announce the first successful community buyout to a gathering of enraptured party faithful in Glasgow in October 2017. To a swelling chorus of applause and delighted cheers, she said:

> And we have started Scotland's modern journey of land reform.
>
> That journey continues today.
>
> In recent months Ulva, an island off the west coast of Mull, has sought permission to [enter into community ownership].
>
> If permission is granted, the residents can get on with raising the money needed.
>
> The Scottish Government has carefully considered the application.

And I am delighted to announce that we have today granted permission to the people of Ulva to bring their island into community ownership.

She could barely get her words out, so loud and delighted were her listeners. The six current inhabitants of the island heard the news that evening and professed themselves as thrilled as the conferees down in Glasgow. The leading lights of the Northwest Mull Community Woodland Company, a local forestry group that would now act as the "community" purchasing the island, had been given due warning of the announcement, and they were wondering how best to raise the £4.25 million at which Ulva had been valued for sale.

The only person who knew nothing of the suddenly changed circumstance was Jamie Howard, Laird of Ulva. He was on the island, checking on his cattle at the time. His daughter telephoned from London to tell him what Nicola Sturgeon had said. He was to be, quite without warning, effectively dispossessed.

The Scottish government, he learned, had stepped in at the last moment. The sale to the helicopter visitor, which Jamie Howard was anticipating would probably go through, was to be canceled. He could barely believe what he was hearing. "I was completely flattened, and hugely disappointed that they could have treated

The tiny Inner Hebridean island of Ulva, with a population of six, and hitherto owned by a single family, was sold in 2018 to a community of locals, under new Scottish laws encouraging wider ownership of the country's land.

us this way," he told *The Scotsman* newspaper. But the following morning his worst fears were confirmed: a letter of pitiless economy arrived, its contents quite unanticipated. It informed the sitting laird that under the rules by which Scotland was now being run, he no longer had the right to sell his island to any of the individuals who had expressed interest. He had to take the island off the market right away. The only entity that under the 2016 Land Reform Act was legally permitted to buy it would be the group of people nearby on Mull who had lately formed themselves into a local *commu-nity*. Ulva was now to be the subject of a community buyout. An independent appraiser would determine what the island was worth and tell the community, which would then be allowed time to raise the money and complete the purchase.

The Howard family were not best pleased. While conventionally wishing the island residents well— all five of them—they issued a statement of cool contempt. The road to the moment of sale—which was consummated on Midsummer's Day 2018, a scant eight months after Ms. Sturgeon's triumphal announcement—had been, they said, "a somewhat dark one. The strong indication is that the driving impulse for this acquisition has been, and is, not so much for the welfare of Ulva and its resident commu-nity but more to satisfy the long-held personal ambi-tions of a relatively few local individuals on the island of Mull; to benefit SNP party politics and prejudices; and to feed media headlines."

The new-formed community did not have to raise much money toward the purchase. The Scottish gov-ernment had set up a land fund specifically designed to help local communities buy land from their land-lords. In the case of Ulva, it donated 95 percent of the £4.25 million purchase price. The remaining and relatively trivial balance of £212,500 could be raised, the government said, by the community members themselves.

They did not have long to wait. Though Ulva has attracted many well-known figures in the past—Sir Walter Scott, Beatrix Potter, Samuel Johnson, and, of course, James Boswell among them—the richest legatee of its former residents, back when it sported some eight hundred people in the mid-nineteenth century,* is an Australian financial giant, the Macquarie Group. The original Macquarie, Major-General Lachlan Macquarie, was born on Ulva, and gained fame for a distinguished army career that culminated in his appointment, in 1809, as the first governor in chief of New South Wales and Tasmania, where he was regarded as a consummate and humanitarian success—"He found a gaol," remarked the *Dictionary*

---

* Their numbers were massively whittled down during the time of the Highland clearances, more notoriously a phenomenon of northern Scotland and discussed in an earlier chapter. The basic thesis of the clearance held that sheep made better and more profitable tenants than crofters, so the major landowners moved humans from their lands—often with brutality and violence—and replaced them with flocks of animals. The displaced crofters headed off either to the cities or, more often, to the Americas—which is what took place for most of the residents of Ulva, now scattered to the four winds.

of National Biography, "and he left a burgeoning colony." His best-known contribution to Australia's early history—and for which to many in the country today he is thought of as "Australia's father"—was his decision to allow all the convicts in what had been established as a penal colony to enjoy, at the end of their sentences, the rights of free men. Provided that they behaved well and generally conducted themselves well, they should be treated "as if they had never transgressed the law."

To honor their founder's connection to Ulva, the Macquarie Group topped up the government's Ulva grant with a healthy infusion of cash, so that the full purchase price could be met. And so, on that Midsummer's Day, once the wire transfers had been made and the checks satisfactorily cleared and an immense pile of paperwork had been signed and sealed and notarized, the Isle of Ulva, the isle of the wolves in Gaelic, passed formally out of the Howard family's possession and into the hands of, among others, the five people who lived on the island. Until that day they had been tenants; on June 21, 2018, they awoke as owners, never liable to pay rent again.

Ulva now has a development plan and a development officer, a young woman well versed in such planning.

She arrived in 2019 from the south of Scotland to come and live in a cottage near the boat landing, thus bringing the population to six. It is hoped the population will rise to twenty by the end of the 2020s—applications have been pouring in, they say—and then rise to fifty once a cluster of affordable housing has been built. The big and none too beautiful mansion house, where the Howards had lived since the matriarch bought the island for a pittance in the 1940s, is to be converted into a visitor center. A herd of some thirty Highland cattle will be imported, and the arable and pasture potential of the 4600 acres will be exploited, the new owners hope, with the aim of becoming profitable by midcentury, if not before. Most recently, there were newspaper reports that the community is trying to win a grant of still more government money, in an effort to get Ulva to pay its way.

Jamie Howard as a result of the sale is acceptably comfortable, financially speaking, lives in Edinburgh and the South of France, and wishes to forget this sorry chapter in his life. He is quoted as saying that he could not now imagine ever going back. Ulva, for him, is history.

What will now happen on Ulva remains to be seen. What has happened thirty miles to the north, on the island of Eigg, however, is by now well established, because a buyout there occurred twenty years before. Eigg was where the notion of community ownership in Scotland got its start.

Actually, not quite true. Technically, there is one very much older example, at a township called Glendale on the northwest corner of the Isle of Skye. In the late nineteenth century the crofters there, legatees like so many in rural Scotland of the injustices of the clearances, found themselves unexpectedly and, considering their impoverishment, cruelly denied two permissions that were crucial to the livelihood: their landlords said they could no longer collect firewood from the beach, and they were told they could no longer graze their cattle on the village green. Led by a firebrand named John Macpherson, they promptly rebelled, and they did so with sufficient vigor, mostly directed at Skye police, that the government directed a gunboat, HMS *Jackal*, to steam into the Pooltiel Loch to try to parlay. William Gladstone, the prime minister of the day, also paid a soothing visit and was much taken with Macpherson's spirit and the strength of his cause. But even so, five Glendale crofters— soon thereafter to be branded "the Martyrs" in the press—were tried for their unruliness and packed off for short prison terms. So great was the consequent negative publicity that once freed they and their fellow community members eventually won their point: new crofting laws were passed. Even though the rebels found those unsatisfactory as well, the situation calmed down when the government itself bought the land, subdivided it into plots for each homestead, and gave them all to the inhabitants, providing them with the security they had long craved.

Since that time, well over a century on, some of the crofters have sold their lots, thereby diluting the notion of the settlements being a true community, as well as the idea of the land thus being community-owned land in a technical sense. Nonetheless, the Glendale crofters now have a place in history. Given that their problems were at last resolved by an eventual government purchase made at the end of 1908, almost ninety years before the buyout of Eigg, they can lay claim to being the first in the movement.*

Until 1997, the island of Eigg—one of the four

---

* And still more pedantically, Eigg is in fact the third, since after Glendale there came the 1923 gift of most of the islands of Lewis and Harris, in the Outer Hebrides, to their inhabitants, by the soap-making millionaire Lord Leverhulme. The complications of this rather difficult transfer of ownership contrasts, however, with the relative simplicity of the later matters relating to Eigg.

Small Isles to the south of Skye, a loose congregation
of geologically spectacular islands that also includes
Rhum, Canna, and the risibly named Muck—had
until 1997 been owned by a variety of individuals
who had, not to put too fine a point on it, rather run
the place into the ground. Like so many Hebridean islands—indeed, like most places in northwest
Scotland—it had once known a measure of isolated
prosperity. At times, under the superintendence of
the Clanranalds, lairds for numberless generations,
there were as many as five hundred residents, growing potatoes and oats and rearing cattle. But come
the nineteenth century and the usual woes began: the
Clanranald family found the burden of island ownership irksome and sold the island for £15,000 to a
doctor named Hugh MacPherson. In short order, he
began to suffer the shocks of the potato famine—or
rather, his tenants did—and like so many landowners, he saw the rising price of wool for textiles as
rendering sheep a far better source of income than
the rents paid by his tenant crofters. So he began his
own cruel program of clearances in 1847, thereby
triggering the all-too-familiar spiral of Scottish de-
population. Nova Scotia, with its similar climate and
topography, was to be the main beneficiary of Eigg

The Isle of Eigg, which for centuries had a succession of
colorful private owners, has since 1997 been owned by its
inhabitants, its subsequent development closely watched as an
early experiment in community-run real estate.

migration, with Barbados, so very different, being
an unexpected runner-up.

After MacPherson sold up in 1893, owners of
varying degrees of incompetence or insouciance
came and went, among them the supposedly liberal
politician Walter Runciman, who is most notable—or
most notorious—for having recommended in 1938
to Prime Minister Neville Chamberlain that the
Nazis be allowed to annex the Sudetenland region

of Czechoslovakia.* When Runciman died in 1949, his sons sold Eigg to the first of a succession of other lesser-known figures—the initial buyer a Welshman whose cattle, soon after he moved in, all died of bracken poisoning. He was followed by a naval commander who wanted to run a school for disabled children, but had few takers; once his money was depleted and he found there was only one pupil remaining in class, he decided to sell—confessing as he did so that far from being a naval officer he was a plain civilian who had only ever commanded a small fire brigade.

It was then, in 1975, that the precipitant figure of the saga purchased the island. He was to be the ninth and penultimate Laird of Eigg: Keith Schellenberg, a mil-

---

* Most relevant here, Runciman spent much of his time in Czechoslovakia in the friendly company of ethnic Germans who complained bitterly about Czech land reforms that they claimed were robbing them of their livelihoods—and thus driving them into the hands of the Nazis. His helpmate at the time was the infamous Stephanie von Hohenlohe, a widely bedded Viennese Jew, the wife of a minor and evidently complaisant Austro-Hungarian prince, and a friend of and suspected spy for Hitler. It is more than probable that stories of Runciman's friendship found their way into dinner table conversation at the Italian-style lodge he built for himself on Eigg, which survives, if somewhat worn out and dilapidated, to this day.

lionaire playboy Yorkshireman. Schellenberg wanted the island so badly that he escaped from being locked inside a castle owned by his second, vengeful wife, by rappeling down the castle walls to a telephone box and from there offering a purchase price of some £274,000 to the erstwhile fire-brigade captain, an offer that was promptly accepted. Like so many of his predecessors—and like so many island owners generally, almost all of them needy and male—Schellenberg intended to create for himself what some psychologists have termed a quasi-uterine utopia, in which he could both rule without challenge and bask in cosseted, quasi-maternal comfort. He promptly turned the island and its small clutch of inhabitants on its head, in so doing inadvertently creating the modern idea of community landownership that so defines—or seeks to define—present-day Scotland.

Schellenberg was by all accounts a handsome, dangerously plausible, devilishly charming, and deeply disagreeable figure. He was well able to afford the firefighter's sale price from the fortune he had already amassed in a string of businesses that involved chemicals, animal feed, and shipbuilding. He was a man of many idle pursuits: he had been variously a bobsled pilot (representing Britain in the 1964 Olympic Games) and a competitor in the kindred winter sport of the luge, as well as a powerboat racer and an inveterate—if vegetarian—partygoer and social climber. He was highly sensitive to criticism and was a feared and capricious litigator. He married four times and he regarded the purchase of Eigg as his opportunity—thanks to Scotland's ancient traditions of feudal landlordism—of creating a near absolute micromonarchy for himself, with his clutch of island tenants under his almost total authority. Almost nothing could be done on his 8000-acre island estate without his permission.

His first intentions seemed honorable enough—with promises of money to be spent in prodigious quantities to right the wrongs of past owners, plans for golf courses and tennis schools and tourism schemes, and exhortations to young people well beyond Scotland to come and help make the island a more vibrant place. But his promises all soon evaporated. It transpired that he had virtually no ready money, having spent great sums on divorce settlements, alimony, and child support. All soon turned out to be smoke and mirrors. He hired and sacked people, abandoned projects in midstream, forgot to pay visiting entertainers for his many extravagant and unaffordable parties—one lordly Scot imported to add glitter and social cachet to one event loudly and drunkenly yelled at the bewildered islanders that they were "the scum of the earth and half-baked socialists." Tenants soon began to claim that their houses, untended by the factor on the often absent Schellenberg's estate, were slowly turning into dilapidated, rat-infested slums. Soon afterward, the island amenities, such as they were—a community hall, most notably—were being closed down.

More of Schellenberg's friends came to the island for ever more raucous parties (with a swastika draped from his mansion balconies on at least one occasion). Schellenberg himself, with a gathering disdain for the islanders, liked to spend exasperated hours racing along Eigg's narrow only road in his 1927 ten-seater Rolls-Royce touring car, scaring people, sheep, and chickens in equal measure and laughing mercilessly as he did so.

If the islanders had any initial fond feelings for this new laird, they vanished in quick time; and their increasingly mulish attitudes of noncooperation and dislike resulted before long in evictions, punishments, and overt threats of violence from teams of burly bailiffs, on call to do Schellenberg's bidding whenever needed. The atmosphere on Eigg deteriorated steadily through the eighties and nineties. Things reached a climax in January 1994, when someone set fire to Schellenberg's hated Rolls-Royce, quite wrecking it and turning it into a blackened shell. The police were called, and a bevy of constables chugged over on a MacBrayne company's ferryboat from Mallaig: none of the islanders would speak to them about what might have happened, other than for each to deny any personal knowledge or responsibility. No clues of value were found; no one was ever called to account for the arson.

Schellenberg promptly lost himself in fury, called the islanders all manner of names—drunken hippies, dropouts, rotten and quite barmy revolutionaries, holders of acid-rock parties—and evicted a group of tenants out of pure spite. The seventy-three islanders (a small number of curious incomers had arrived in previous years), who were by now well-known in the British press—there were few better pegs for a story than an attack on a Rolls-Royce owned by an Olympian with a German name hidden away on a remote Scottish island—set to thinking how best they might be able to buy out Schellenberg and demand that he be on his way. Large local organizations—the Scottish Wildlife Trust, the Highland Council, and Highlands and Islands Enterprise among them—pledged support and in time possibly funds to help create a body, a community-based body, that could perhaps buy the island out from under Schellenberg's increasingly autocratic rule.

To avoid a confrontation and to head off a growing assault on his already tattered reputation—and possibly to cock a snook at the islanders once again—

Schellenberg suddenly sold the island to an eccentric sun-worshipping German artist and soi-disant professor named Gotthilf Christian Eckhard Oesterle, whose chosen nom de plume, Maruma, he had seen reflected in a puddle of water in Abu Dhabi (or else Geneva—accounts vary). He seldom came to the island other than by helicopter, and in any case he had formidable financial problems of his own. When the islanders and their various supporters told him that they had formed the Isle of Eigg Heritage Trust some years before and had now raised enough money to buy the island from him, he agreed and, with a sigh of very evident relief, signed the relevant paperwork. Oesterle headed back home to deal with the publicity problems in Germany that developed when it turned out he had never been a professor of anything, and owed £300,000 to a German clothing exporter.

Accordingly, the trust—the community, as it were, now it was wrapped up in legal form—became on June 12, 1997, and after the payment of £1.5 million to the creditors' lawyers, the official owners of Eigg. The days of lairds—of which there had been ten, Clanranald to Maruma—were now formally and finally over. The island and its 8000 acres, dominated by a high ridge of obsidian visible from miles around and known as the Sgurr, was now owned principally by its inhabitants. Now at last, with the laird gone, the old Scottish system of feudal land tenure, so unfair to so many, so favored for a few—with some of the more impossibly micromanaging previous lairds specifying, for example, which if any of the island's many varieties of seaweed could be eaten by their tenants—was to be ended forever.

There was a blaze of publicity in the months and years immediately following the ownership change, not all of it good. The island population increased quite rapidly, from a low of forty people before Schellenberg, by way of seventy-odd during his tenure and then getting on for a hundred once he had gone. Eigg's demographic mix changed profoundly too, from an originally Gaelic-speaking, Hebridean near monoculture to an omnium-gatherum of newcomers from well beyond, including a small number of progressively minded incomers from—*quelle horreur!*—the English Midlands. There was expected to be a consequent clash between Gael and Sassenach, between progressivism and conservatism, between youth and retiree. The press waited, hoping for news.

A German magazine was among the first, writing an unkind essay on the theme of Haight-Ashbury meets the Hebrides. The incoming community members—dismissively caricatured as Green Party–

supporting, bearded, ponytailed, unwashed, sandal-wearing, bohemian, and decidedly noncorporate types, people who fled the cities and its competitiveness for a more sustainable and cooperative life—were to the German readers members of a tribe of classically disaffected misfits, the kind of people who had once probably lived in Tibet—as if—and had crude signs hand-painted on their rusting minivans. Clashes were predicted, disaster widely anticipated, the notion of the community ownership of land—this being essentially the first such experiment—widely dismissed as an impertinence by admirers of the old ways, and with withering scorn.

But in fact most visitors today find that the communities are getting on with one another tolerably well, all doing their best to make Eigg work without the burdensome presence of a lordly landowner. One hundred and ten people live on the island at the time of this writing, with some infants having been born on Eigg to newcomers. All manner of physical betterment is visible. There is homegrown electricity now, courtesy of a trio of wind generators by a farm just below the cliffs of the Sgurr, a small hydroturbine on a particularly turbulent stream, and a cluster of solar panels. There is telephone service and a modest amount of internet bandwidth. There is a small school, an even smaller brewery, a general store, a café, and a taxi driver. The twelve-bedroom Italianate mansion built by the Runcimans back in the thirties belongs now to a couple, Norah Barnes and Bob Wallace, who have set up a center for sustainability studies; and if there is a certain well-worn tick-all-the-boxes predictability about it—Iyengar yoga, meditation, wellness, eco-building, earthship construction—there can be little doubt that the lodge is being put to rather better use than when a giant swastika was employed as part of the decorations for extravagant weekend house parties. There are three children of the Barnes-Wallace family,* so they can feel secure in the knowledge that a doctor now visits the island from Skye each Tuesday if the weather allows. A veterinarian will call if needed, as will a dentist. Otherwise the islanders are left much to their own devices, initiative, and intestinal fortitude.

Government subsidies have helped finance many of the island's various recent improvements, and taxpayers from time to time complain. And not all like the newly made community nor its inhabitants—a

---

* The name might sound familiar to those interested in the history of war. Barnes Wallis, his surname differently spelled, was the heroic inventor of the so-called bouncing bomb, famously used for the RAF dam-busting operation against the German Ruhr in May 1943.

delivery-truck driver complained that whenever he came to Eigg he found the islanders "lolling in the grass, drinking, doing drugs, generally being lazy" and never helping him load or unload his cargo. Moreover, he said, warming to his theme, the community members were "always disagreeing with one another, talking endlessly, doing nothing, spending taxpayers' money, dreaming, wasting everyone's time."

Far better, he continued, that islands were owned by individuals, men who had vision and competence—and then, just as he spoke, and entirely by chance, the CalMac ferry to Mallaig, for which we were waiting, swept up to the quayside, bearing as its sole passenger an island owner of the very kind the disgruntled driver favored. He was the laird of the tiny neighboring island of Muck, a craggy highlander named Lawrence McEwen, known the length of Scotland for his kindly eccentricity and for the abiding loveliness of the island that his family has owned since the Great War. He could scarcely cut more different a figure than Keith Schellenberg—modest, self-effacing, weather-beaten, as tough as old boots, dressed like a hobo and beloved by the forty-odd tenants who live on Muck in conditions of modest respectability, with no store, no postbox, electricity only very recently—and with no apparent wish for it to be otherwise. The mix of farming, fishing, timber harvesting, and sporting lets—deer hunting, partridge shooting, and the like—by which McEwen keeps the Isle of Muck afloat, just, leaves few islanders unhappy with the arrangement. In theory McEwen runs his domain just as he wishes; but since his wishes seem to be broadly in concert with those of his tenants, and since he seems little better off than any of them, most appear content to keep matters as they are.

By all accounts, the social engineering of Scotland's once feudal landownership system is gathering pace. Nearly six hundred pieces of territory are now community-owned, their previous owners flush with cash and many of them living far away, many happily dispossessed of the responsibility that rightly comes with ownership. More than half a million acres—2 percent of Scotland's land surface—is now owned communally. The new arrangements have as expected brought withering criticism from established landowners. They insist that community members are poor managers—they talk without acting, have little experience of the realities of farming, neglect their animals, plant ill-suited crops, allow fences to break and buildings to rot, shoot animals and leave them lying dead in the heather, take and waste so much government money that they trigger waves of envy in their nonsubsidized neighbors, and describe islands like Eigg

Twenty years ago, this blasted mountain, the famous Sgurr of Eigg, was owned by just one person, as were all the island acres spread out beneath. Today a hundred islanders own it communally, reflecting a profound change that has come upon modern Scotland and may soon spread far beyond.

as becoming like "gone-wrong American communes from the 1960s." Moreover, and to the wholesale dismay of one particular admirer of the old, traditional ways, newly imposed taxes on shooting have recently become a main source of revenue for the Scottish Land Fund—the very fund that is used to finance community buyouts. "We owners are thus being taxed to fund the means for buying ourselves out. We are asked, in other words, to become the agents of our own destruction."

The changes, he said despondently, were surely all political. That land reform around the world has a political component to it should come as no surprise. But that politics is the *only* true consideration in the reapportioning of land in Scotland—as seems to be the case on Eigg, on Ulva, and now in hundreds of other recently changed ownerships elsewhere—is being greeted by many with some dismay, insofar as it seems to them to challenge the very essence of what Scotland long has been all about. Whether the new and supposedly more equitably disposed Scotland prospers as a socially healthy and economically vibrant entity, with its land widely dispersed throughout its community, has still not been fully tested and is still not fully known.

3
Bringing Africa Home

We fear Africa because, when we leave it alone, it works.

—PATRICK MARNHAM,
Fantastic Invasion (1980)

Oh god!" the countess remarks in tones of languid dismay. She is standing on the east terrace of her farm, gazing across the Great Rift Valley toward the rising African sun. "Oh god! Not another fucking beautiful day."

She is by birth an American, once Alice Silverthorne of Buffalo, more recently the Countess de Janzé. She is a dissolute, promiscuous murderess, a real person captured memorably on film by Sarah Miles in the 1987 British melodrama *White Mischief.* Her dawn moment says all too much about the languor and luxury of western colonial behavior, endured in much of the rest of the world, but perhaps most egregiously in the near limitless tracts of Africa.

For a while at least, the archly hedonistic life of those who lived in the highlands of Kenya as members of the so-called Happy Valley Set was well-nigh unimaginable to most in the rest of the otherwise war-torn world of 1940. And if all was in the end just an illusion, then it was most grandly so. It was a time of manners that in many ways represented the apogee of European imperial rule in Africa: it was a period of rule soon to crumble away, to be replaced by other manners, other behaviors, with many of them in truth little more edifying than this.

The time had begun at the close of the eighteenth century, in ways that all empires seem to begin—with what some might have supposed was full of promise, marked by episodes of heroism and honor, by science and investigation and serious and sympathetic interest. Periods in history that are marked by exploration and survey, by botanical and biological and anthropological inquiry, seem so often to be the precursors of conquest, of cruelty, and of mercantile greed. And this was as true in Africa as it already had been elsewhere. After the initial sub-Saharan explorations had been completed, once heroic figures like Mungo Park, like Sir Richard Burton and John Speke, and, of course, like Henry Stanley and his long sought prize, Doctor Livingstone, had all come home, written their books, colored their maps, given their lectures, and collected their medals, that madcap period which historians would call *the scramble for Africa* began, and it did so in particular earnest.

As usual the missionaries went out first, eradicating as best they could what they saw as the spiritual impoverishment of animism and witch doctoring, and bringing as many natives as possible under the ecclesiastical authority of Rome or, better still, of Canterbury. In the missionaries' wake, and once the more exotic indigenous passions—suspected cannibalism chief among them—had been subdued, the merchant venturers and their guardian soldiers set out, all of them in search of fortune.

These figures—the lately reviled über-imperialist Cecil Rhodes most notably—had no interest in souls, but sported only a frenzied hunger for African land and, most particularly, for the treasures that they believed lay beneath. Their appetite for territory knew no bounds, quite literally. For sixty frantic years begin-

THE RHODES COLOSSUS
STRIDING FROM CAPE TOWN TO CAIRO.

Linley Sambourne's 1892 *Punch* cartoon of Cecil Rhodes bestriding Africa from the Cape to Cairo has long offered an iconic representation of imperial adventuring across the continent and the wholesale seizing of its lands.

ning in the last half of the nineteenth century, thousands of settlers and fortune hunters sailed from chilly northern seaports bent on carving out new countries, on establishing new cities, on displacing inconveniently sited peoples, and on establishing mechanisms of governance and administration that paid little heed to the settled customs of those who remained.

An entire ancient continent, stretching from the placid Mediterranean waters off Cairo and Tangier down to the icy chill of the Southern Ocean beyond the Cape of Good Hope, was casually and ruthlessly partitioned. It was divided up, Nigeria to Natal, for the ultimate benefit, in all too many instances, of languid and louche inhabitants like those of the cool uplands of Kenya Colony, where the British came to rule, and to play.

The British were far from being the only ones to blame, of course. European greed for African land was a universal ill, with the Belgians in the Congo every bit as rapacious as the Germans in Namibia and Tanganyika, with the Portuguese in Angola and Mozambique quite as distempered as the Dutch in the Transvaal, Spain's west African conquistadores just as vengeful as those on the far side of the Atlantic, the French and the Italians perhaps saucing their conquering misdeeds with cuisines that long outlasted them (Abidjan's creperies being legendary). Even the Swedes, the Norwegians, and the Danes, peoples customarily revered for reasonableness and quiet courtesy, did little of lasting good during their very much earlier and brief sojourns on the continent.

But perhaps because English was to become the lingua franca of most of educated Africa, it is the British who remain most closely associated with the imperial phase of the continent's long story. And so it is perhaps appropriate that it was a speech made by the British prime minister Harold Macmillan, to the South African

Parliament in February 1960, that would mark, and for all of Europe's other colonists as well, the beginning of the end of this imperial era.

"The wind of change is blowing through this continent," the patrician statesman declaimed, anticipating the headlines for London's next morning newspapers. "And whether we like it or not, this growth of national consciousness is a political fact.

"We must all accept it as a fact, and our national policies must take account of it."

Africans were about to get their countries back. The anti-imperial rebellions had started decades before: the Mahdi's precursive stand in Sudan long ago, the Ashanti rising in the Gold Coast at the turn of the century, the Matabele and Mashona troubles in Rhodesia, the insurrectionary habits of the Zulu in Natal, the Mau Mau insurgency in 1950s Kenya, the Ibo in Nigerian Biafra, the rising of Frelimo in Mozambique. Civil unrest and extreme violence and open warfare was so much the leitmotif of the previous hundred years of Africa's history that it was widely accepted that only a wholesale European withdrawal could allow for a lasting peace. All the more so since, in theory at least, some of those who would now be getting their countries back were going to get their lands back as well. The Happy Valley Set and their like would—in many cases, but by no means all—be persuaded, obliged, or forced to hand back their estates and their farms to the peoples of Kenya from whom they had been taken years before—whether they liked it or not. And what would happen in Kenya would happen in Uganda too—and in Tanganyika, Bechuanaland, Nyasaland, Somaliland, the Belgian Congo, and South-West Africa and in myriad other corners of the continent that would now be cleansed of foreign influence and governance, and whose institutions and peoples would be remade such that foreign domination—by which, essentially, one meant white domination—would never be allowed to occur again.

And so, one by one, the chartered jets flew south with important European personages aboard, and these personages, often dripping with diamonds and magnificence, hosted final formal dinners at Government Houses, made declamatory speeches full of flourish and earnest promise, signed parchment documents of retrocession and release, and presided the following morning over military march-pasts and many-gunned salutes and the lowering of the European flags and the raising of new and colorful African banners, and there were aerodrome handshakes filmed against the background whine of jet engines and then final farewell waves from outside the aircraft door and after takeoff maybe the dab of an eye with a handkerchief, as beneath the plane the panorama of the great green landscape of Africa—with her forests and her red murrum roads and her clouds of circling egrets—was at first made vague by wisps of cloud, and then vanished clear away as the plane gained altitude, after which it was time to make a course back north to the chilly airports of Europe, and to leave Africa to find her own way home.

Such passing-out ceremonies punctuated the sixties, the seventies, the eighties, with an almost dizzying abundance. Once the drone of the last departing aircraft had faded, and the interlopers had all left, the maps they had drawn and the borders they had created—it is worth noting that there had never been any entity called Nigeria, for instance, until Sir Frederick Lugard had designed it,* no Rhodesia until Cecil Rhodes had struck a mining deal with Chief Lobengula of the local Matabele and had the resulting concessioned territory named after himself—were put to use, in most cases, as the geopolitical framework for the construction of the new Africa.

Not a few country names were changed, however:

---

* And had his wife, in a letter written to *The Times*, name the colony after the great Saharan river that flowed into the Atlantic Ocean through the locally British-run jungles.

the world saw the birth of locally run nations called Namibia, Botswana, Zambia, Zimbabwe, Tanzania, Malawi, Zaire. And not a few periods of strife then had to be endured, and various tyrants opposed or deposed as time went by, and as new governments tried shakily to settle themselves into place.

It would be idle to say that the nation-building efforts of new Africa have been everywhere a consummate success—except that the continent does now enjoy the knowledge, if such brings comfort, that those who run Africa today are no longer strangers from beyond Africa's coasts, but are the African tribes of ancient times, admitted to full flourish once again. And so the continent is once again a place of Ashanti and Shona, of Herero and San, of Toro and Ndebele and Zulu and Swazi and Tuareg, of Hausa, Masai, Kikuyu, and Matabele, the poetry of their hundreds of names—and languages and dialects and appearances and traditions and customs and clothing and gods—adding to the barely believable complexity of the land in and on which they live.

Which makes the essential notion of *land reform*—a concept so central to the idea of nation building, and yet in itself very much a white man's construct—so wretchedly complicated. Reforming the ownership of land anywhere on the planet is a trying task at the best of times, as these pages have tried to demonstrate. But when the challenges posed by broken colonial treaties and land-grabbing avarice and long established custom and institutional resistance are mixed together with the competing claims of scores, hundreds, and maybe thousands of tribes—Zambia alone has seventy-two, from the swamp-dwelling Ambo in the west to the plantain-farming Yombe of the dry center—then tradition-based complexity risks turning into social catastrophe.

Whether discussing tribal Africans or New Zealand Maori or Arizonan Hopi or the sea people of Polynesia, the very concept of landowndership is to some cultures both puzzling and alien. To a Matabele chief interviewed recently in Zimbabwe, the idea of buying land was just insane: *You might as well buy the wind*, he retorted. Reformers may try until they are blue in the face to suggest that the simple fact of ownership brings with it a vast catalogue of the benefits of capitalism— and yet the chief will sigh and thank the visitor for his time and will insist that the traditional system of land tenure, whereby he, just like his predecessors for generations past, grants the rights of tenancy to his tribesmen, works perfectly well, and all are content. But, the reformer will protest—the insecurity? All are content, the chief insists, and the matter is closed.

Tribal land tenure, coupled with slash-and-burn agriculture and nomadic behaviors—the last more common among those tribes living closer to the continent's western and northern deserts, where water supplies are fitful and settled farming is more risky—still dominates the landscape of much of sub-Saharan Africa. The situation of the white settlers was almost always formalized, however: Cecil Rhodes, believing he had the perfect right to hand out land that he declared was nominally owned by Queen Victoria, gave parcels of it to the Rhodesian pioneers, with formally written and printed title documents. At the same time he herded the tribal peoples into reserves, immense acreages of elephant-grass countryside that he saw as perfectly suited to their relatively unsophisticated agricultural practices, and for which they had no particular need for documentation.

The contrast was stark: on the one hand were the white farmers, legatees of the former colonial apparatus, knowing that they held in their bank vaults, the original paperwork of deeds and the associated bundles of legal rights; and on the other hand those black natives who, despite nominally having their kinfolk now running the new-made country, in so many cases lacked any evidence of legal tenure—and, in many cases, lacked any appreciation of any need to have it.

Such reality has lain at the heart of Africa's land reform problem for the last half century, at least.

And all too often consequent nightmares have dominated the headlines. Zimbabwe's Robert Mugabe is the figure most commonly associated with the more disagreeable attempts to change Africa's balance of land-ownership.

He seemed to take an initially progressive stance. Land, he first agreed, could be reallocated voluntarily, with the money to do so offered as a conciliatory gesture by the former British colonial government. A fund set up in London would pay market prices to willing white farmers, and their purchased land would then be distributed freely to Zimbabwe's landless peasants. By 2000, some 40 percent of land previously owned by whites had been transferred into black hands. Except that it turned out not, as had been promised, into the hands of landless peasants. In all too many cases, it had gone instead to the hands of Mugabe's political cronies—a discovery that prompted London to freeze its fund and publicly to deplore what Britain saw as the president's breach of trust.

After which matters went from bad to worse—and the country was plunged into an anarchic spasm of anti-white violence and confiscation. Agents of Mugabe's ruling party, styling themselves "war veterans"—tough young men who claimed, often baselessly, to be former fighters in the long-drawn-out Rhodesian Bush War that ended with Zimbabwean independence in 1980—invaded farm after farm, with invariably humiliating, distressing, and sometimes lethal consequences for the owners. Only a scattering of white farmers now remain in the country; the land has indeed been duly redistributed (with the late Robert Mugabe's family somehow coming to own fifteen of the country's largest farms). However, the agricultural bounty that had long underpinned one of the healthiest economies in southern Africa is no more; not only has Zimbabwe been reduced in recent years to a state of beggary, but most think of the nation as an international pariah. Criticism of Mugabe's self-evidently failed land policies is also regarded by many, however, as being fueled by an innate racism; and it has proven difficult to this day significantly to elevate the argument, and thus to win any hope of bettering the situation.

Elsewhere on the continent, the redistribution of the immense expanses of potentially fertile countryside has met with similar problems, though seldom marked by such violence. In South Africa in 1998, Nelson Mandela enthusiastically championed the same kind of "willing-seller, willing-buyer" redistribution program that the British had supported in Zimbabwe; but it worked only

During the thirty-year rule of Zimbabwean leader Robert Mugabe, there were many episodes of the forced confiscation of land from white farmers by so-called war veterans bent on what the government saw as retributive justice.

fitfully, and since his death in 2013 there has been a clamor for a more robust program of restitution: *expropriation without compensation* being the phrase of today. The legacy of past injustices haunts South Africa: most notably, the Native Land Act of 1913, which restricted Black people to living in just 13 percent of the colonial territory, and the reality that during the apartheid years some 85 percent of the land was owned by the tiny white minority. In 2017 a survey showed that nearly three-quarters of private farmland in South

Africa was owned by whites, who make up just 9 percent of the population. There is much noise and unrest in today's South Africa, but little of moment is being accomplished: Mandela's saintly idealism and hopes for peaceable redistribution of his country's three hundred million acres is slowly slipping into the archival memory.

Five hundred million acres of African land currently remains uncultivated: and yet six hundred million African people—almost half of the continent's population—exist below the World Bank's recently upgraded $1.90-a-day poverty line. The authors of almost all studies seem to throw up their hands after studying this seeming paradox—of so much land and so little wealth. They apportion blame in familiar ways: corruption, laziness, a lack of proper documentation, bureaucratic incompetence, the legacy of white colonialism—all are responsible for a seemingly intractable problem. Other countries beyond Africa that were once seized with long-term rural poverty—Brazil, Indonesia, Argentina, China, and, even to an extent, poor, hapless, and benighted India—have made immense improvements in their rural lot; but Africa, which superintends fully half the planet's uncultivated land, seems pinioned in an immobile and intractable mire.

In the uplands of today's Kenya, farming still continues apace. The Happy Valley Set may have long ago vanished; Alice, the notorious Countess de Janzé, died at the age of forty-one, wrecked and ruined by drugs, by alcohol, tobacco, and unsuitable men. In the closing days of empire there was something of an official realization that the White Highland lifestyle that spawned such creatures as she was wholly inappropriate, was not consonant with the new imperial image. An ordinance was passed in 1961 that saw the formal end of an extraordinary colonial edict that had once reserved the Highlands for white Europeans exclusively, and which had the local Maasai and Kikuyu all put out and sent into the woodlands below.

The independent Kenyan government of today nonetheless accepts, perhaps through gritted teeth, that some white farmers from the Cotswolds, Scotland, and the Yorkshire wolds may still have something to offer by working the cool uplands around the towns of Naivasha, Gilgil, and Nakuru, on properties all the way to Eldoret. So young men with clipped accents, cavalry twill, and an education in agriculture at Cirencester still journey out by plane to Nairobi and then head north—though no longer on the old Lunatic Line, the lion-infested railway that used to take passengers up to Uganda—to work in the hills, growing tea and coffee, sisal and chrysanthemums.

The dawn still comes up across the Great Rift as always, of course—except that now it is not only white expatriates standing on their farm terraces who may view it. Some modest reforms have been managed here, and have taken hold, perhaps setting an example to all the continent beyond. So the endless succession of beautiful African days that so captivated the many colonials eight years ago are as a result deservedly available these days to African men and women too, the delights to be experienced by Kenyans standing on farm terraces that look out over long stretches of fertile landscapes of which these days they are the rightful owners, and at last.

4

Aliens in Wonderland

That first full gaze up the opposite height! Can I ever forget it? . . . the modicum of moonlight . . . gave to that precipice a vagueness of outline, an indefinite vastness, a ghostly or weird spirituality. Had the mountain spoken to me in audible voice, or begun to lean over with the purpose of burying me beneath its crushing mass, I should hardly have been surprised.

—HORACE GREELEY,
*An Overland Journey from New York
to San Francisco* (1860)

Conservation, especially in light of today's warming and ever more polluted planet, is near universally seen as an admirable practice. After all, who could possibly find any fault with what the OED defines as "the preservation, protection or restoration of the natural environment and of wildlife"? The wild outdoors is surely a sacred space, best kept pristine and pure—a place where, as the Scots-born environmentalist John Muir once remarked, "all the world seems a church, and the mountains altars."

Except that there are many places of inestimable and churchly beauty in the world in which the wildlife includes not just elk, deer, bears, eagles, bluebirds, salmon, and alligators—but human beings.

People, in other words, native people who have lived in these wild places for thousands of years and for whom the land and its beauty—a beauty that is precious to all visitors fortunate enough to come see it— can fairly be said to be theirs, to belong to them, to be the very community of which they have long been a vital component part.

The Yosemite Valley of northern California is one such place, and people did indeed live there for thousands of years. But these people have long since been cleared off it, shooed away by soldiers, no less. They were sent away for the supposed ultimate benefit of the millions who now come to Yosemite National Park to marvel at the wonders of Nature. Those visitors come to worship before the sublime works of the Almighty God, which in Yosemite are writ large across three-quarters of a million acres of Sierras foothills marked with gigantic granite cliffs, snow-covered peaks, and great torrents and cascades of icy rushing water.

If this language is inflated, that is because pioneers of environmentalism like John Muir wished it to be so. The high-minded Calvinist approach to land, of which

The saintly reputation of Scotsman John Muir, an early conservationist and cofounder of the Sierra Club, has been somewhat tarnished in recent years because of his white supremacist views and evident disdain for Native Americans and Black people.

Muir was the chief spokesman and is now the most hallowed patriarch, was central to America's nineteenth-century conservation movement. Its core belief held that land was sacred and God-given, and that man was by contrast sinful, unclean, and unworthy. Muir, who was a confirmed eugenicist, a man dismissive of the laziness of those he termed "Sambos," and yet who went on to found the Sierra Club, with its millions of liberal-minded members today—saw the natural world as the "conductor of divinity." and, indeed, that Nature was quite literally at one with God. And that a place of such sublime beauty as Yosemite—which he championed in two famously lyrical late-nineteenth-century essays for *Century* magazine, winning it popularity clear across a nation eager to explore—was a heaven-sent benison, and needed to be preserved at all costs from the depredations of humankind. Especially, as he saw it, from the local Native Americans, whose "dirty and irregular life" was a stain on the otherwise ethereal beauty of the valley in which they had chosen to live.

Mankind was a malignancy that needed to be kept at arms' length from such beauteous creations of geology, biology, and botany. Virtuous bodies with names like Earth First! have lately sprung up to denounce the very existence of millions of defiling human beings who are now living but who should preferably, in the opinions of such ardent earth lovers, be properly dead. In a remark that has a peculiar resonance to those with vivid memories of the 2020 viral pandemic, the royal consort the Duke of Edinburgh once reflected that should he ever be reincarnated he hoped it would be as a deadly pathogen, one that might help solve human overpopulation and the harm it persisted on visiting on the natural, God-given landscape.

It follows that adherents of such beliefs—in America most particularly—have long been keen to restrict and control all human activity within those areas deemed most beautiful, the most needing to be cherished. Humans may under certain circumstances visit and contemplate the majesty of God's empty wilderness—but empty they must be, and if not empty, then they must be emptied. This was the doctrine behind the passing in 1864 of the Yosemite Land Grant Act, signed into law by President Lincoln. That legislation preserved for all time and made sacrosanct the valley of the Merced River—which flowed down westward from the high summits of the Sierra—and is considered the first step in creating today's redoubtable system of similarly protected national parks.

What helped make the Yosemite Valley relatively easy for the federal government to protect in 1864 was the fact that almost no human beings were living there, nor had been for the previous fourteen years. They had been cleared away ahead of time by a figure famous in California's pioneer history, James Savage, who had come out west on one of the famous wagon trains from his home state of Illinois in the 1840s and had eventually pitched in with hundreds of others at the time in searching for gold. Savage, however, first picked up some military experience by volunteering in 1846 to fight—though he seems never to have taken part in combat—in the Mexican-American War. Once he was discharged, he decided to set up trading posts in the Central Valley. Most notably, he established a post that he called Agua Fria beside a river that had been named by Spanish explorers for its springtime abundance of monarch butterflies—quite a nuisance to their soldiers, it seems—as the Mariposa Creek.

It was from here at Mariposa that Savage ventured out, leading a party of state-sanctioned and heavily armed militiamen, to put down some "pesky" Indians who were frustrating the activities of local gold prospectors—and in the process he "discovered," as white men used to say, the stunning landscape of the Yosemite Valley. And began the process of clearing it of its longtime resident native people.

Longtime indeed. Archaeological records suggest

that communities of Miwok and Mono people of California and Nevada had existed there for at least eight thousand years. The Miwok (the word in their language means "people," so tautologies abound) inhabited lands to the west of the Sierra ridgeline, while the Mono, linguistically and genetically rather different, lived in the dry rain-shadow deserts to the east. The Mono were nomadic, as desert people generally are. The Miwok—distributed statewide and making up a linguistically linked group that is divided by western scholars according to geography into Bay, Coast, Lake, Plains, and three kinds of Sierra Miwok, the northern, central, and southern—were very much more settled. They built villages, had ceremonial roundhouses, constructed sweat lodges, granaries, and cabins, cleared land for growing corn, gathered nuts, and wove baskets. They were a sophisticated and gentle people— ill served by the thousands of white newcomers who poured in the Californian foothills in search of gold.

These miners and prospectors—the famous forty-niners, named for the year of their first arrival—were wholly disdainful of the Indians, regarding them as primitive, unclean, savage. They saw it as their right as new Californians to enter Miwok lands with impunity, there to plunder this or that lode, or else to employ vast water cannon to try to blast nuggets from the alluvial cliffs in the lower reaches of the Sierra rivers. As it happened, the places where most gold was to be found were on the very same acres where wild game was most abundant—just below the Sierras' winter snowline— and so were the most thickly populated with Miwok villages. It is scarcely surprising that from time to time the Native Americans raised objection to the strangers' ruin of their traditional lands, and that from time to time they fought back; entered trading posts, mine shafts, adits, and the sheds where the hoses were stored; and, by doing damage, vented their frustration or rage. One can hardly blame them: this had been their land for generations, and now stampedes of crude outsiders were despoiling it. If on occasion they fought back, then who in all conscience could object?

But the settlers did object, and after his Agua Fria property was attacked—and three of his workers killed—James Savage petitioned the governor of the newly formed state of California governor, a young Ohio immigrant named John McDougal, for permission to establish a volunteer militia to deal with the problem. A force of two hundred mercenaries, the Mariposa Battalion, was promptly raised in early 1851, and set out through the Sierra foothills both to conduct punitive raids and to try to negotiate with the enraged central Sierra Miwoks, who were doing their level best to make life as miserable for the forty-niners as they were making it for them.

Savage himself—tall, blond, always wearing a red shirt (the better to impress the Indians, he said)—conducted some of the negotiations, having a considerable aptitude for the native language. A federal commission also tried to stage talks with the Indians, but neither the commission nor Savage himself made much headway. And so in March 1851, exasperated by the indecision, Savage's attacks got under way. The levies rode out, each of the battalion's three companies assigned to wreak havoc in one of three main valleys. Scores of Indian villages were burned, hundreds of tribesmen and women were forced away from their ancestral farms, hunting grounds, and homes, and the Miwoks were directed toward the hot and dry Central Valley, to reservations that had been readied for them, far away from the contentious goldfields in the foothills.

It was during one of the first of these punitive expeditions, in late March, that one of the battalion's raiding parties ventured up along the valley of the Merced River in search of miscreants. Attached as surgeon was Lafayette Houghton Bunnell, a twenty-seven-year-old doctor's son from Rochester, New

York, who had taken the opportunity to explore the valley because, two years beforehand, he had been nearby and declared the scenery to be memorable. As he later wrote, he had seen "the stupendous peaks of the Sierra Nevada . . . in the distance an immense cliff, I looked upon this awe-inspiring column with wonder and admiration . . . whenever an opportunity afforded, I made inquiries concerning the scenery of that locality. But few of the miners had noticed any of its special peculiarities."

What Bunnell had spotted, gleaming like a beacon in the evening sunshine, and what he now rode toward along the valley that early spring day in 1851, was the star attraction of today's Yosemite National Park. He had seen the sheer shining face of El Capitan, brilliant and almost reflectively glasslike against the backdrop of snow-covered peaks to the east. As the party rode into the valley, deeper and deeper between the towering walls—shooing away Indians as they met them, directing them to break down their cabins and the sweat lodges and head back downriver into the flatlands to the west—the sights Bunnell had previously glimpsed from afar, came steadily into full view. El Capitan, for sure. Then the Half Dome. The Bridal Veil Falls. Cathedral Peak. At every bend in the river there was something memorable, fantastic.

Everyone in the party, Bunnell most especially, was stunned by what they were seeing. Landscapes like this—with such soaring peaks, vast and sheer rockfaces, grand waterfalls, snow-covered ranges, flower-filled meadows, forest groves, rocks of a satin smoothness warming in the sunshine, giant sequoias, glaciers, deep blue lakes, the scent of pine needles everywhere—had never been seen before, anywhere. The Mariposa Battalion had stumbled into a sublime and heavenly place, and the world needed to know about it.

Such miners as had passed through the valley had evinced no interest beyond the amount of gold to be panned from the Merced—precious little beyond a few flecks, as it turned out. So "it would be better to give it an Indian name," wrote Bunnell later, "than to import a strange and inexpressive one; that the name of the tribe who had occupied it would be more appropriate than any I had heard suggested. I then proposed that we give the valley the name of Yo-sem-i-ty, as it was suggestive, euphonious, and certainly American; that by so doing, the name of the tribe of Indians which we had met leaving their homes in this valley, perhaps never to return, would be perpetuated."

Melancholia was to be found in the grout of this otherwise anodyne explanation: "the tribe who *had* occupied it," "leaving their homes," "perhaps never to return." For that was what Savage and his company ensured: that the sedate village life of these harmless and congenial Miwok was now essentially over, reservation life beckoned, and the Yosemite Valley was forever destined to be pristine, preserved inviolate, and all but free of sinful humanity. Like Adam and Eve, mankind had been evicted from Paradise. Yosemite was ready to be prepared for the altar of conservation for the good of the people of America—in those days, largely white people—to behold.

The Miwok Indians, who had long lived in and around Yosemite, were by and large turfed out of their traditional lands to make way for the onrush of tourists, though some were retained as attractions and entertainment for the visitors.

Thirteen years later, on June 30, 1864, Abraham Lincoln took time off from his Civil War responsibilities to sign the two-paragraph Yosemite Land Grant Act, handing the lands officially over to the custody and care of the state of California for "preservation and public enjoyment." Within months, photographs of the quite extraordinary landscape were appearing all across the country. The public swiftly got the message. Travelers arrived in large numbers, by horse, in traps. In carriages. In time, cars started to make their way to the outwash of the valley, and promoters started to advertise the marvels within. Roads were built. Then parking lots—first for horse-drawn vehicles, later for automobiles. Tunnels were carved through the giant sequoia trees—one mighty and ancient specimen fell soon thereafter; another was weakened by runoff from all the cement nearby. Concessions were opened. Hotels. Souvenirs. A few Miwok were brought back from their reservations to work as servants and guides and to display, if they still knew, the crafts and manners of those who had once lived there. John Muir became involved, distressed at the now excessive public use—by sinful humanity, as his beliefs suggested; he campaigned that the land be taken out of California's control and placed it in the hands of the nation, made into a national park. This it was in 1890, eighteen years after Yellowstone in Montana had been made the inaugural member of the National Park Service.

There were 6000 Miwok living in the Sierra at the beginning of the gold rush, before the raising of the Mariposa Battalion. At the time President Lincoln signed the Yosemite Land Grant, there were just 100 Miwok remaining. Today, matters have improved somewhat: all told, the Miwok of the central Sierra—the so-called Yosemite Miwok—amount to around 150 people, living in a reservation that abuts the northern side of the park. There they run the Black Oak Casino, as state law allows.

Five million visitors come to Yosemite every year, bringing $379 million in local economic benefit. Some of that bounty is no doubt spent at the casino, though John Muir would presumably disapprove.

5

Trust Is Everything

We abuse land because we regard it as a commodity belonging to us. When we see land as a community to which we belong, we may begin to use it with love and respect.

—ALDO LEOPOLD,
A Sand County Almanac (1948)

Should one wish to illustrate the meaning of the word *placid*, then the lower reaches of the Charles River in Boston, viewed on a crisp and breezeless evening in the fall, would serve just perfectly. The water here is wide, flat, and mirror still, the river's seaward flow near imperceptible. The skylines—Harvard and

MIT in Cambridge on the north side, Boston University and the downtown skyscrapers on the other—provide the familiar backdrop for the college rowing eights gliding past or, as dusk approaches, the single sculls that pass by, serene and swanlike in their solitude, leaving the briefest imprint on the definitively placid surface of the stream.

Gazing from the riverbank here, it is easy to forget that the Charles has a wilder side to it, that it has a source, is ruffled by rapids, is squeezed between cliffs, that there is a place where the river and anyone beside it can be quite unaware that it will in time become wide and placid. But seventy winding river miles away from Boston—twenty as the crow flies—there is such a place, a gorge, a ravine, a deep chasm that back in colonial times the settlers called the Gates of the Charles, since that was where the river appeared to rise. Nowadays there are golf courses nearby, an old insane asylum, a clutch of suburban tract housing, a women's liberal arts college, a goodly number of farms, quarries that once provided stone for Boston, and the headquarters of the Bose Corporation, makers of powerful loudspeakers and much sought after earphones. All this activity flourishes to greater or lesser degrees outside, in this much populated quarter of eastern Massachusetts. But inside, within the stone walls of what is now officially called the Rocky Narrows Reserve, there is nothing. Nothing except for nature—nothing but forest, meadows, rocks, and water, just as it always was, preserved intact and inviolate for the community at large.

Rocky Narrows, with its 274 acres of quite unspoiled Massachusetts landscape, has the distinction of being the oldest surviving possession of the oldest privately run and not-for-profit land conservation trust in the United States. It was given for the public good in 1897 to a body that had been established six years previously as the Trustees of the Public Reservations (the word *Public* was dropped in 1954). It had been founded by a landscape architect named Charles Eliot, a man with the vision of having the community at large become the trustee-owners of the land around them, for all to enjoy and for none to keep private. He started a movement that, while evolving over the decades into something as fearsomely complex as only the American legal profession can make such things, has lately become a powerful force for returning the land to the ownership of all. Helping, in other words, to make land communally owned, and doing so organically and apolitically, and not enforced artificially, as in the cases of Eigg and Ulva and elsewhere in the more decidedly political atmosphere of modern-day Scotland.

There are essentially four tiers to this revolutionary new means of owning and stewarding land in the United States today. There are trusts that operate on a statewide level; those that work solely within a county; there are town and village land trusts; and down among the weeds there are the community land trusts, the much smaller-scale groups that are somewhat less involved with landscape and prettiness and leisure, more with affordability and want and social need.

The Trustees of the Reservations in Massachusetts got off to a flying start with the gift of the Rocky Narrows. An eminent Brahmin—Augustus Hemenway, the heir to an enormous South American silver-mine fortune who is memorialized today by the immense eponymous gymnasium he gave to his alma mater, Harvard College—made the original bequest. The transfer was handled by a friend and early landscape-gardener employer of Eliot, Frederick Law Olmsted, one of the designers of New York's Central Park. Not only is the tract physically beautiful, with its cliffs and hemlock stands and pinewoods and pastures, but it also serves as a powerful reminder of the first and bloodiest of all the wars between white settlers and Native Americans—in this case, the local Wampanoag Indians and their charismatic leader, Metacomet, also known as King Philip. The gorge was a strategically important defensive line for the Indians, though as it turned out not so effective, since the Wampanoag were fought to the death in a series of brutal battles—King Philip was himself horribly butchered—and then were all but annihilated as a tribe by the end of the seventeenth century.

The Trustees were the first in the United States—and almost certainly the first in any country—to establish the principle behind the privately funded land conservation movement. It relies on the altruism of citizens who can be encouraged to do one of two things: to give away tracts of their own land that they wish to have preserved in perpetuity: or to give what in Massachusetts are called conservation restrictions (in the forty-nine other states, they are conservation easements), which allow the land to be used only in certain carefully prescribed ways, while the owner retains ownership. He can give, sell, or will his land as he pleases—just so long as the restriction remains in place forever—and with the same restriction or easement passing down through all successive owners as well. This sometimes limits the sale price—not everyone wants to buy land that, say, positively encourages people to roam across it—but the proponents of such schemes think the greater good is served, and thus so be it.

This last provision is probably what had made this particular conservation movement such a success. Almost four hundred conservation restrictions have thus far been given by landowners across Massachusetts, involving some 21,000 acres. One of the properties involved, in the Berkshire hills at the far western end of the state, is named Questing. The story of its 400 acres, which lie in the prettily sprawling cluster of villages that constitute the town of New Marlborough, is typical of the slow and very positive land revolution that seems now to be spreading across America's hitherto very vulnerable countryside, where developers of shopping malls and trailer parks once seemed to run wild.

Questing is a simple enough reserve, far less pretentious than its name suggests. There is room for half a dozen cars to park outside a farm gate. An old forest road then leads uphill through glades of hardwood, ash and oak and cherry, until the woods open up into a meadow, seventeen acres of grassland alive in the summertime with wafts of butterflies and dragonflies. Beyond is forest—hemlock, white pine, yellow birch, red oak, maple—peppered with the ruined cellar-foundations of houses built three centuries or more ago, and with seasonal streams and vernal pools and old stone walls, and looping around it all just one rough path, marked only with blue-paint blazes on trees a hundred yards or so apart, making it less difficult to get lost.

Birdsong is everywhere; there are mushrooms and

salamanders, mosses and ferns, and every so often the sudden red flash of a frightened deer, scurrying out of sight. The term *forest bathing* may be overworked these days, but to be quite immersed in a wood of such unassuming coziness is positively rejuvenating for the spirit. All the more so when, after an hour or so of seeing no evidence of anything modern, no people, nothing but wild nature and some well-worn reliquaries of New England settlement history, you come back out into the meadow again and behold!, there is a wooden bench at which you may rest and take in the view, and silently thank the Trustees for caring so admirably for a tract of community land, gratefully given and lovingly tended for the benefit of all.

This piece of land was left to the Trustees by a pharmacologist named Robert Lehman, who died nearby in 1996. He had enjoyed some success in New York City as the inventor of a pharmacopoeia of chemicals that helped manage a variety of sicknesses and had bought the ruin of a farm built in Victorian times by a pair of brothers named Leffingwell, who had both died in farming accidents—one killed by a falling barn beam, the other by the kick of a horse—and whose family then left for the flatter terrain of central Indiana. Lehman and his wife spent decades restoring their old farmhouse, using it as a second home for a while and then moving in full-time once the restoration had made it habitable. They purchased a variety of small parcels of land surrounding it over the years, but always with the intention of granting almost all of the acreage to the community by way of the Trustees, once they had no further use for it. Once, in other words, they had passed on.

And so the community owns it and uses it today, with gratitude. Since it is managed at long distance—the Trustees' headquarters is a hundred miles away, near Boston—the acreage is left more or less untended, although the current owner of the Leffingwell house set on the few acres retained privately under the terms of the Lehman grant looks after the trails privately, as best he can. And since Massachusetts is a very small state, less than three hours' drive from end to end, there is a feeling of intimacy about the arrangement—a once locally owned parcel of property is being cared for by more-or-less-local trustees, who know a thing or two about the vegetation, forestry, and climate that is peculiar to these hills.

In addition to state-centered land trusts like this, there are trusts that operate in America on a county level, and, in a more recent development, there are those that operate in a more granular fashion still, down at the town and village level. It is these last that seem to be enjoying the greatest current success. By the latest count, there are about thirteen hundred small-town land trusts across America, protecting and preserving through the principle of local private ownership—or by using perpetual conservation restrictions—a total of some ten million acres. Land that can never, ever be developed.

This, it would seem, is the near ideal situation: local land, offered as gifts by locals, bought with funds raised by locals, sculpted and made accessible by locals, managed and now tended to by locals and now used freely and generally by locals. And yet all of it has been created informally, organically, from the ground up, if the pun may be allowed. And mercifully free from politics, ideology, and argument.

This is by no means Eigg in America: it is not the result of doctrinal pronouncements, political speeches, and the scoring of partisan points. This is community landownership pure and simple. By all accounts, it is an arrangement that appears to work well, offering to the local public the undisputed benefits of health and exercise and peace and quiet and serenity that all need from time to time, and in the manner that the noted ecologist Aldo Leopold had so famously wished for. This is land at its very best—owned by all, used by all, for the good of all.

And finally, one level still further down, at a truly grass-roots level, there is the idea of the community land trust. This is an idea bruited very much longer ago, and born not in America but in India; it concerns itself much less with landholdings than with housing, with cities rather than countryside, and with mitigating poverty rather than with mandating serenity and pleasure.

Vinoba Bhave, who was a friend and disciple of the Mahatma Gandhi, first established the idea of such a voluntary land reform movement in the early 1950s. Like his mentor, he was an ascetic and given to spectacular, nonviolent demonstrations of his political beliefs. For example, in order to publicize what he saw as the need to break up existing feudal systems in India—the situation whereby titled princes, nizams, maharajahs, as well as zamindars, *walis*, and nawabs owned huge swathes of countryside, and yet millions were left wholly landless—he did a classically Gandhian thing: he picked up a stick and a mendicant's begging bowl and set out to walk thousands of miles, barefoot or in *chappals*, across the entirety of his country.

Tall, thin, bearded, and at the time a good sixty years old, he wore only a homespun *lungi* from a *khadi* shop. His walks were eventually followed by thousands, such that a vast snakelike procession of land reform-

The Gandhian disciple Vinoba Bhave walked across India in the 1950s, urging landowners each to donate a sixth of their holdings to the local people as part of what he called the Bhoodan movement. It fizzled.

ists swept across the land, its leader bent on persuading all owners he encountered or alongside whose lands he walked to give up some of their territory—a sixth of it, Bhave usually suggested—and to do so voluntarily and out of a sense of simple decency. And in their hundreds, they did, and for a while India became a repository of a considerable acreage of what was to be called Bhoodan Land—Gifted Land—on which hitherto landless peasants could live, and which they could till and seed and raise crops and on which they could feed a water buffalo or two, and thereby secure an entry-level position in a path to eventual prosperity.

Sadly, and being India, the Bhoodan movement eventually fizzled. Bhave is generally regarded today as a saintly figure, a prophet more or less without honor in his own country, and he has appeared on an Indian postage stamp. But very few of the pieces of land he once so persuasively pried from the hands of various zamindars and nawabs belong to landless *Dalits* anymore. Most have somehow become subsumed into other estates or, where they once stood close to cities, have become unrecognizable components of Indian suburban sprawl, the details of their ownership mysteriously now passed into the hands of members of the fast-growing Indian middle class.

Yet some of Bhave's ideas—that landowners should behave charitably and give away parcels of land for the specific purpose of helping to alleviate poverty—have long survived him. Moreover, they spread very much more successfully outside India, and nowadays the notion of city land being gifted for the purpose of helping impoverished urban dwellers has advanced somewhat. Many hundreds of community land trusts have sprung up in recent years, particularly in Britain and the

United States. The parcels of land involved—generally, they are reasonably small; this is a movement less interested in landscape than in building plots—are acquired in much the same way Bhave sought to acquire his land—as gifts freely offered up by right-minded people. Some governments—regional governments, in the case of modern Scotland, state governments in one or two progressive-minded administrations elsewhere—help pay for buyouts. There are many different models. But once the land parcels are transferred, local land trusts administer them, generally in perpetuity, and see that inexpensive, attainable housing is built on them, either for sale or for rent. The land beneath always remains the property of the trust; the improvement upon the land owned or rented is otherwise the responsibility of the occupier.

The rural American south saw in the 1970s the first extensive rollout of an idea that had much in common with the kibbutz scheme in Israel, as well as with the philosophical approach of the Jewish National Fund, which accumulated land in Palestine and offered ultra-long leases to worthy landless applicants. A somewhat similarly established community trust was first established in the American south in the seventies—"A non-profit organization to hold land in perpetual trust for the permanent use of rural communities," according to its brochures. It raised sufficient funds to buy a 5000-acre farm in southwestern Georgia, and farmers developed it for twenty years. This experiment ultimately failed, but it set the movement in motion, generating enthusiasm among those in the community who followed such land-economy philosophers as Henry George,* the celebrated inventor of the radical idea of the land value tax.

The Georgia model spread, slowly at first. It took root next in Cincinnati, then most significantly in three counties in Vermont on the shores of Lake Champlain—where the local Champlain Housing Trust managed to acquire a sufficient acreage on which to build 2,300

---

* In his wildly popular polemic *Progress and Poverty* (1879) the writer and political economist Henry George proposed that all taxes be abolished except for that applied to the potential rental value of unimproved land—what many have since described as the "perfect" tax. Henry George had an enormous following in America in the late nineteenth century, and his funeral in New York in 1897 drew the largest crowds ever then seen in the city. Few present-day economists could ever imagine such a send-off; yet his legacy has never found favor in any advanced country on earth, despite George's firm belief, supported by great numbers of thinkers, that since land is a near immutable gift of Nature, logic dictates that it is the only possession that should be subject to taxation. *Too difficult to administer* is the usual argument against.

apartments, more than four hundred single-family houses, and fifteen commercial buildings—giving work and shelter to some six thousand people, who otherwise would have struggled to find places to sleep. Of the 250 community land trusts in the United States, that in Vermont is by far the largest, the flagship.

In Britain the popularity of the movement is growing at similar speed—at the time of this writing there were slightly more trusts than in the United States: 255 of them in England, Wales, and Scotland. In Scotland trusts have permanent ownership of half a million acres, a figure that includes the aforementioned islands of Eigg and Ulva, in which the transfer of ownership from private to community involved some direction from government, rather than free-enterprise goodness, which appears to the hallmark of the movement across in America.

It is one thing to take a progressive idea like the community land trust and employ it in a progressively minded state like Vermont, quite another to try it out in one of the more broken cities of America's Rust Belt. But since 2008 a group of local activists has been experimenting with a raft of such radical ideas in the once spectacularly forlorn city of Cleveland, on the Ohio shores of Lake Erie.

Until lately, no better example of urban failure could

The journalist and tax reform advocate Henry George was a hugely popular figure in late-nineteenth-century America with his 1879 book *Progress and Poverty* assuming magisterial authority. He proposed and advertised the consolidation of all revenue-raising efforts into a single Land Value Tax.

be found than Cleveland—a two-hundred-year-old port that once stood at a nexus of roadways and rail lines, a city built on steel mills and heavy industry—which during the Second World War particularly was one of the busiest industrial cities in the country. But after the war the demography altered: there was white flight to the suburbs, migrations of unemployed men from the Deep South, redlining by mortgage companies, closures of factories. And the pollution was infamous: the Cuyahoga River was so laden with oil and flammable chemicals that quite regularly it burst into flames, and one such river fire in 1969 was so bad it damaged and nearly destroyed two iron railroad bridges. Race riots broke out with dismal frequency. Cleveland in the 1970s was a city in serious decline, with an aspect of hopelessness about it that seemed irreversible and incurable.

After thirty years of decay and desuetude, a group of seers with progressive ideas arrived on the scene, with plans for maybe curing the city ills. In 2008, an experiment known as the Evergreen Cooperative was launched, with the intention of making sweeping reforms to the employment and housing situation, to the use of local land, and to improving the urban environment—and it was established with the enthusiastic cooperation of a weary and relieved Cleveland city government and the local hospitals and health-care companies that had lately built offices there. By all accounts the initiatives seem to be having some positive effects. The chronic population decline has flattened. New building has increased. There are new parks, bright new schools, and—at least until the 2020 pandemic struck—an increasing tax base.

And the Cuyahoga River has become clean enough today to support some fish, if not necessarily pure enough to drink. Where once its waters oozed dankly between rotting wharves and ruined warehouses, now it sparkles in the sunshine, and families push children in strollers along pathways beside its banks. Cleveland is by no means out of trouble, but the kind of land reform ideas that were born in Georgia in the 1970s seem to be a key component of a raft of changes now being effected in one of America's most grimly injured cities. The old place at long last is showing signs of beginning to heal.

Sharing and distributing land more liberally, as it appears to be being shared here and in Cincinnati and in rural Vermont and Georgia, can do unanticipated good. If properly and fairly apportioned, land can be the key to so many possibilities, all of them for the general benefit of those of us who live and work and have our being upon it.

Epilogue
Yet Now the Land
Is Drowning

The near universally held belief that underpins almost all of the dealings with land that I have described in this book can be summed up thus: *land is the only thing on this earth that lasts*. Some in the recent past have waxed lyrical on the theme: land is the only thing truly worth working for, worth living for, even worth dying for—and that is because its enduring nature is absolutely undeniable, is unchallengeable. The lasting nature of the solid surface of the world is axiomatic; it should be taken for granted, is as guaranteed as the setting and rising of the sun.

And yet, as we now know, this is not so. In relatively recent times much new land-related knowledge has been gained, and as a result much has changed in our beliefs. Since 1965, for instance, we have come to know that continents are far more plastic than they look and that their land, once supposedly immobile, does indeed move, and that the plates to which this land is all bolted shift and jostle and plunge and collide one with the other in a ceaseless ballet of wandering rock. No one of intelligence alive today can doubt that continents are anything but fixed. We scoff at the naïveté and mulish ignorance of our forefathers who ever supposed otherwise. *Eppur*, as was famously remarked four centuries ago about the world entire, *si muove*.

Similarly with the gospel that spoke then of the immutability of the land itself. Maybe, yes, some grudges would now say, it can be permitted to move somewhat—to deny what science has now shown to be true would be foolhardy indeed. But the quartermaster's accounts still hold true, surely? The simple existential fact of the presence of 37 billion acres of the planet's exposed surface is as sound a basis for surety as is the existence of gravity, or of mathematics. Two plus two is four. g runs at thirty-two feet per second per second. These are constants. Pi is a constant. Boyle's law speaks of constancy. And similarly, land. Land endures. Land stays put. Land lasts. Its shape may change. It may now move a little, yes. But its totality remains the same, aeon by aeon forevermore.

Except that since 2005, and a thicket of alarming

news from those who study the strangeness of the climate, we have lately come to accept the truth of the very opposite.

Land is decidedly not staying put. Land is in fact withering away. The seas that surround the land are rising, and they are rising fast. They are doing so because the world is getting warmer. The glacier ice and the Greenland ice and the Antarctic continental ice are all melting and pouring into the oceans, which are themselves becoming warmer, and their warming waters are swelling in volume. Tides are getting higher and storms are getting more frequent and surges are occurring in places where there never were surges before—and in sum, the land is under threat like never before in human existence.

The land is drowning. It is slowly but steadily shrinking away, like a balloon with a leaky valve. And sensible humanity is being urged all around the planet to move away from the seas' shorelines, and in due course shift to safer and higher ground.

Thus far, the loss of land has been all but imperceptible. Between 1996 and 2011, for example, the East Coast of the United States, from Maine to Florida, lost just 13,000 of its acres to the clawing waters of the Atlantic Ocean. One would suppose this to be a tiny amount, scarcely worthy of mention. A year later, however, came Hurricane Sandy, which hit New York City square on and then lingered: suddenly, rising-water nightmares and flood scenarios that New Yorkers had never imagined became an urgent reality. The previously imperceptible became the immediately catastrophic. Enormous areas of Brooklyn, Queens, and Staten Island streetscapes were swept underwater, torrents of salt water poured into the always open gates of the subway system, road tunnels became truck-drowning floodways. Rising ocean levels became overnight a New York certainty, and as a result new sea

The East Coast of the United States is particular vulnerable to rising sea levels, and many acres are already in 2020 being chewed away by oceanic storm surges and tidal erosion.

walls, flood barriers, diversion canals, and pumping stations are being completed or built as the scientists warn and the politicians allow. The trivial-seeming inundations of the recent past are recognized for what they were truly were: auguries of a certain kind of global doom.

The fate of the planetary land depends today on a feedback loop, a self-fulfilling prophecy. The steady melting of the once pure-white polar ice sheets has the simple effect of decreasing the planet's reflectivity, its *albedo*. A nonreflective earth allows more solar heat to reach down and do its damage, melting more and making the now off-white world ever more gray. And the more gray, so still more heat is allowed to be absorbed and so still more gray all becomes, with the result that the circle deepens and widens and the melting accelerates—and the sea continues to rise up and up along the tide tables, and the storms become ever more numerous and dangerous, and land, more land, much more land, becomes swamp, becomes flood prone, slides underwater, then vanishes itself into the sea and is settled into becoming sea and finally being sea, forever.

The figures now suggest that this kind of flooding will accelerate, in and around New York and elsewhere, and more urgently in those poorer, lower-lying countries that cannot afford to build barriers to help save themselves. This is a problem manifestly more urgent for those countries that will lose land that is thickly peopled, and by people who do not have the funds or the choices that might enable them to withstand, mitigate, or ameliorate the unstoppable forces of nature.

Whole island groups will go. Out in the mid-Pacific, the Marshall Islands, which include Bikini and Eniwetok, where the first atom bombs were tested, will soon become green with shallow water and in later years blue when the water inundates them wholly. The vast-spread pattern of islands that make up today's Republic of Kiribati, which being positioned both astride the International Date Line and the equator manages, uniquely, to exist in both the Northern and Southern Hemispheres and to experience today and yesterday, summer and winter, all at the same time, will go under too. In French Polynesia, Claude Levi-Strauss's *Triste Tropique* will become reality, Paul Gauguin's *Nevermore* similarly, as rising waters erode old island landscapes and change islanders' lives forever.

Oceanic expansion will be truly global in its consequences. This rising of the waters will wash away mudflats in Bangladesh and West Bengal and Thailand and Burma, as well as eroding the estuary of the Yangtze in Shanghai, deepening the waters in New

Orleans, flooding the docks in Oakland, London, and Valparaiso, turning the Fens and sands of East Anglia into open water.

For a while, and for most people, the land losses will still be imperceptible, will be thought of as little more than omens, worries for the long term and faraway. Populations will adjust. Coastlines will be nibbled away, not gnawed at greedily. Big cities farther inland—Kolkata, Dhaka, Nanjing, Rangoon, Lima—will, if reluctantly, receive most of the nearby displaced. New Zealand is alive to the local problem and has said formally that it will accept some Pacific Islanders as climate-change refugees. Other countries may follow suit.

It will be some long while before the 37 billion acres of the world are diminished by any serious fraction. It will be very many decades at least before as much as a single billion acres, for instance, could be thought of as being at immediate risk of loss. A billion acres is a combination of the areas of India and Pakistan together, is almost twice the combined areas of Alaska and Texas. It will be decades, centuries most probably, before land on that scale will be lost to the world. The coastal erosion experienced since 1996 along the Atlantic coast of North America amounts to a loss only of an acreage the size of San Marino, or of half of the island of Anguilla. Maybe by the end of the twenty-first century the world's land surface will have shrunk by an area equivalent to, say, Belgium. The coastal lands are indeed withering, and they will not last. But they are withering slowly enough for many—if not necessarily most—to avoid the contemplation of making any specific plans to leave.

Notably, however, the world's major private land-owners do not own or dominate coastlines to any significant degree. They may have their seaside mansions in New York's Hamptons or in Britain's Sandbanks or in Cap Ferrat or on the Queensland coast—which the flood-prediction maps all show as inevitable victims of the coming sea. But they have covered their bets—most probably unintentionally—in almost all cases by acquiring their long-term landholdings far away from the sea and at very much higher, flood-proof altitudes. The John Malones and the Gina Rineharts and the Ted Turners and the countless Scottish dukes own most of their acres back up in the hills and the heather, among the Rockies and the iron-rich Sierra interiors and the appositely named Scottish Highlands, with few of their holdings vulnerable to any elements of shrinkage.

Although maybe, just maybe, the simple spreading knowledge of the vulnerability of land generally may prompt some of them to indulge in moments of sober reflection. It might give pause to some of the less greedy landowners—and maybe even the greedier ones too—to wonder just why the need to own so much when, in philosophical terms, the ownership of such an unownable entity as land means so little. Of course, the purely economic argument still lies in favor of ownership increases—in strictly monetary terms, a shrinking asset becomes ever more valuable, commands an ever higher premium as it becomes more rare. And yet: might the very fact of land's newly realized impermanence not suggest to some that this could be the time to consider what has for so very long been well beyond consideration—the notion of sharing land, rather than merely owning it, outright?

This would be an easier matter to contemplate if the vulnerable asset were something more readily tangible—if it were corn, oxygen, water, or fish—if the commodity in question, which had hitherto been available in limitless amounts, was suddenly seen to be depleting, and at a time when the world's people had urgent need of it. In such a case even the most heartless would surely agree: rather than hoard it for himself, why not share it out, offer it for the betterment of the needy? And though the depletion rate of the world's land is by no means an urgent business, the fact that it is depleting at all may give some the opportunity to wonder: What if I might share it out, and allow those who truly need it to have use or possession before it all passes beneath the rising waters?

Naïvely idealistic, one might reasonably say—and yet the mantra of those from whom we took the lands in the first place was, all too often, just that. Sharing land is by no means a revolutionary idea. The aboriginal Australians, the Maori, the Canadian First Nations populations, the Inuit who inhabit the high latitudes from Siberia to Alaska and back again, the Aztec, the Incas, the North American Indians—to all and each of these, land was a commodity so precious and so life-giving that it was indeed to be shared by all, and owned by none.

Examples of such a kindly, philosophical approach to the world's surface can found around the planet. Australian *indigenes* see the Earth as their mother, in constant need of care, admiration, gratitude, and respect. In west Africa the elders of the Ashanti tribe declare that land "belongs to a vast family of whom many are dead, a few are living, and a countless host are still unborn." To the Goldi people of far eastern Russia every morsel of land and all that grows from it, be it animal or plant, is possessed of a spirit that is to be worshipped and respected. To harm the earth in any way is sinful and unworthy. Akira Kurosawa's great film *Dersu Uzala* famously depicts the dichotomy between such animist beliefs and the cruder ways of a Russian survey team that arrives in the forest to make maps: the movie's stirring narrative—with giant windstorms, torrential floods, tiger attacks, and all manner of other characteristically Kurosawan drama—depicts the ways of the eponymous Dersu, a local hunter, ways as very evidently the more noble, and enduring.

And in the United States too, representations of the true value of land—of its spiritual value, well beyond its mere monetary worth—can be found occasionally in public declarations of Native Americans. Most famously, perhaps, are the sentiments that were so vividly expressed by the great Chief Sealth—of what is now Seattle—when in 1854 he was persuaded to enter into a treaty with white settlers. The newcomers wanted the chief to cede 2.5 million acres of his choicest coastal land—the irony is inescapable, given coastal land's new fragility—which at the time was the hunting and gathering and settlement area of his people, as well as their burial ground.

Sealth, tall and thundering in his manner—and a Catholic to boot, converted in 1848 by French missionaries—is said to have stood at what would be the Puget Sound waterside one blustery spring day in mid-March 1854 and offered a great oration to his gathered people. Over the years many versions of this speech—or this letter; some say he had it written down and mailed to President Franklin Pierce—have been published, each of them different and yet each asserted to be true. Most are florid, implausible, horribly embellished. What follows is still too wordy, but it seems to scholars among the more reliable (if, indeed, there ever was a speech or a letter made or sent at all—some doubt it). But the words of this ring fairly true, and the sentiments certainly reflect the views of most Native Americans, from Miwok to Mohican, from Choctaw to Cherokee, all of them at the time bowing in forced subservience to their inevitable conquest by the power and greed of the advancing white man.

This is what the wise old man—he was sixty-eight at the time; he would live a further dozen years—is supposed to have said. He spoke of his acceptance of the settlers' offer, but he did so with evident apprehension and regret. Sealth could never have imagined the great iron and glass towers that would eventually rise on his land, particularly in the city that would be named in his honor. Nor could he have supposed that men like Jeff Bezos and Bill Gates, men with such very different values to his own, would become the local tribal leaders of their times. "The President in Washington," Sealth declared,

The Suquamish leader Chief Sealth, after whom the City of Seattle is named, spoke with reportedly great eloquence of his regret and melancholy in handing over his lands for white settlement in 1854. His speech, much romanticized, exists in many versions.

sends word that he wishes to buy our land. But how can you buy or sell the sky? Buy or sell the land? The idea is strange to us. If we do not own the freshness of the air and the sparkle of the water, how can you buy them?

Every part of the earth is sacred to my people. Every shining pine needle, every sandy shore, every mist in the dark woods, every meadow, every humming insect. All are holy in the memory and experience of my people.

We know the sap which courses through the trees as we know the blood that courses through our veins. We are part of the earth and it is part of us. The perfumed flowers are our sisters. The bear, the deer, the great eagle, these are our brothers. The rocky crests, the dew in the meadow, the body heat of the pony, and man all belong to the same family.

The shining water that moves in the streams and rivers is not just water, but the blood of our ancestors. If we sell you our land, you must remember that it is sacred. Each glossy reflection in the clear waters of the lakes tells of events and memories in the life of my people. The water's murmur is the voice of my father's father.

The rivers are our brothers. They quench our thirst. They carry our canoes and feed our children. So you must give the rivers the kindness that you would give any brother.

If we sell you our land, remember that the air is precious to us, that the air shares its spirit with all the life that it supports. The wind that gave our grandfather his first breath also received his last sigh. The wind also gives our children the spirit of life. So if we sell our land, you must keep it apart and sacred, as a place where man can go to taste the wind that is sweetened by the meadow flowers.

Will you teach your children what we have taught our children? That the earth is our mother? What befalls the earth befalls all the sons of the earth.

This we know: the earth does not belong to man, man belongs to the earth. All things are connected like the blood that unites us all. Man did not weave the web of life, he is merely a strand in it. Whatever he does to the web, he does to himself.

One thing we know: our God is also your God. The earth is precious to him and to harm the earth is to heap contempt on its creator.

Your destiny is a mystery to us. What will happen when the buffalo are all slaughtered? The wild horses tamed? What will happen when the secret corners of the forest are heavy with the scent of many men and the view of the ripe hills is blotted with talking wires? Where will the thicket be? Gone! Where will the eagle be? Gone! And what is to say goodbye to the swift pony and then hunt? The end of living and the beginning of survival. When the last red man has vanished with this wilderness, and his memory is only the shadow of a cloud moving across the prairie, will these shores and forests still be here? Will there be any of the spirit of my people left?

We love this earth as a newborn loves its mother's heartbeat. So, if we sell you our land, love it as we have loved it. Care for it, as we have cared for it. Hold in your mind the memory of the land as it is when you receive it. Preserve the land for all children, and love it, as God loves us.

As we are part of the land, you too are part of the land. This earth is precious to us. It is also precious to you.

One thing we know—there is only one God. No man, be he Red man or White man, can be apart. We are all brothers after all.

Thirty-some years after Chief Sealth offered these words, and halfway across the world, a man of similar age and standing was to pose a related question, one that haunts us still.

In 1886 Leo Tolstoy, with both *War and Peace* and *Anna Karenina* successfully published, and his literary reputation assured—offered up a brief parable under the title "How Much Land Does a Man Need?"

The question of numbers—of *how much?*—is especially pertinent in the United States. This is a country

Thomas Jefferson's view of the yeoman farmer, who with his few acres and dedication to hard work would transform the fledgling America, has long been idealized in cartoons such as this, from 1845.

where so much land exists, and yet where so little of it is owned by the country's native people—who never believed in private ownership anyway—while so much of it is possessed by what can only be described as white newcomers. Might a man who presently owns two million acres of land that once belonged to Iroquois or Sioux ever dare to contemplate how much does he really think he needs, to live a decent and fulfilling life? Just what might be the right number?

There are some hints from history. Thomas Jefferson's belief that the sturdy yeoman farmer was the key to the creation of a successful republic had in mind that a man and his family, if all were put to work, might make a living from 50 acres. If the soil was thick and the harvests good, he could maybe handle a little more, maybe as many as 200. Many years later the first of the Homestead Acts thought it meet and proper to distribute the country's public lands, in newly formed and unpeopled states and territories like Nebraska, Oklahoma, and the Dakotas, in parcels of 160 acres or so. The figure represents a full quarter of a *section*, this last being a uniquely American areal measurement, its boundaries established along lines extending from the so-called point of beginning in East Liverpool, Ohio, the place from which all west-of-the-Ohio America was first surveyed and measured. Some lands deemed more difficult to work—if covered by forest, for example— were handed out in full sections, 640 acres for each worthy applicant. Then again, in the period of Reconstruction that followed the Civil War, freed slaves were initially offered the chance of acquiring the famous "forty acres and a mule," an offer made by the U.S. army in part to ensure that slaves did not take immediate possession of the plantation lands on which they had worked. This offer was rescinded in short order, however, a decision that caused a degree of disappointment and bitterness that lingers to this day. The disparity between the amount of land owned today by Blacks and by whites—the average Black household holding assets of no more than 8 percent of those owned by the median white household, land being a central component of those assets—is an enduring legacy that contributes to the country's racial disharmony.

In sum, then, successive early American laws and policies have indicated that, in order to achieve success and contentment, a citizen once needed between 40 and 640 acres of improvable land. No doubt similar numbers could be quite easily determined in other countries—likely it would be a smaller number in the much more populated and congested Europe; and a rather larger number in Russia, where the steppe extends without seeming end from sunrise across to sundown, a vast and thinly populated expanse of rich, black earth, yearning for improvement. This what Leo Tolstoy saw from the study in his great estate at Yasnaya Polyana: his seemingly simple rhetorical question sought to determine how much land might any man need. It sought also to delve into the mystery of why any man could think of himself as actually owning a piece of what, in essence, eternally belongs to Nature.

Leo Tolstoy, wealthy and aristocratic, nonetheless had settled and critical views about landownership in Russia—and the avarice that impelled so many to accumulate vast holdings on the steppe.

The protagonist in his story is a peasant named Pakhom, a man who, while seeming outwardly content with his respectably impoverished lot, holds tight to one particular grievance. As he declares to his household one day, "I don't have enough land. Give me enough of that and I'd fear no one—not even the Devil himself." But as it happens the Devil is listening to this rumination, and he decides, in a typically Tolstoyan literary device, that he'll now have "a little game" with poor Pakhom. "I shall see that you have plenty of land," he says with a diabolical snigger, "and that way I'll get you in my clutches!"

Soon thereafter, some nearby land comes on the market, from a local widow with whom Pakhom has long been on friendly terms. She owns three hundred acres: Pakhom thinks he might be able to afford a small fraction of it. So he scrimps and saves, selling a foal and some bees, borrowing from his brother-in-law, sending a son to get wages in advance for menial work—and finally comes up with enough cash to pay for thirty acres of partly wooded land—not ideal, but good enough to farm. Now, at last, he is a landowner—and within a year he becomes a successful one, by dint of hard work and intelligence. "Whenever he rode out to plough the land which was now his for ever, or to inspect his young corn or meadows, he was filled with joy . . . he lived a landowner's life, and he was happy."

The Devil, however, has work to do. Before long, and for a variety of reasons, Pakhom begins to get too big for his boots—angering his neighbors, falling out with former friends, behaving badly, and ultimately leaving for another village, for a larger estate of some hundred acres of land, then repeating the experience and leaving yet again, and this time coming into the ownership of no less than thirteen hundred acres, well on his way now to becoming a serious, big-time land-owner. By now he is a man who can perhaps one day dream of marching with the Muscovite aristocracy, of becoming a member of the Russian landed gentry. Greed consumes him. The Devil is well at work.

And it is at this point that Pakhom hears a rumor—of an exceptionally innocent and craven group of country people called the Bashkirs who live way down in the south of the country. He hears tell of them that "there is so much land that you couldn't walk round it all in a year. It all belongs to the Bashkirs. Yes, the people there are as stupid as sheep and you can get the land off them for practically nothing." And so Pakhom, salivating at the prospect, sells up once again and leaves for the south. After traveling for seven full days, laden with gifts of tea and vodka, he finally comes to the Bashkir settlement—finding them to be, as foretold, a people who are "very ignorant, knew no Russian, but were kindly."

They turn out indeed to be willing to sell their land—but seeing in Pakhom his very evident eagerness to own a great deal of it, they make him an unusual offer. They will sell him land just as anyone else might sell land, with title documents exchanged, deeds made clear, and so on. But as the village elder explains, the price in rubles will be determined not by the acre, but by the *day*. It will cost a thousand rubles a day.

Pakhom protests that he doesn't understand. "A day? How many acres would that be?"

The elder responds: "We don't reckon your way. We sell by the day. However much you can walk round in a day will be yours. And the price is a thousand rubles a day."

To Pakhom, with the Devil of avarice evidently whispering in his ear, the deal seems unbeatable. He is young, he is fit, the weather is fair, the countryside pleasing to the eye and hardly challenging to walk through. So before the appointed day he slept well; he rises before dawn and begins his walk. Thirty-five miles he should make by sunset, easily. He needs only to return to his starting point, where the Bashkirs and their elder will be waiting for him with a pen and the contract paper ready to be signed, and he will hand over the cash and be the owner of a mighty new estate.

As he walks, so he marks out his passage, turning slightly leftward every so often to keep going in a large circle. He leaves markers every three miles or so. It gets warmer, he begins to shed clothing. His boots starts to hurt. His feet are cut by the sharp grass and the rocks. What he had supposed might be easy is proving to be a challenge. But the more he walks, the faster he goes, the more territory he encloses with his widening gyre, the larger his estate will eventually be. At noon he reaches the point where he will turn back, but is by now farther from the starting point than he imagined, so he begins to trot, then lope, then run full tilt to be sure of getting back in time. And he almost doesn't make it. By the time he sees the waiting Bashkirs, the sun is sliding down across the horizon, and he has to sprint with all his remaining strength, the Devil willing him ever onward, on to the waiting contract, on to achieving the fulfillment of his dream to own so vast an estate of land.

And at last, in the gathering dark of dusk, he lurches forward and falls face-down before the Bashkir elder—who cries out his congratulations, and declares with delight to the effect that *my word, that's a lot of land you've earned yourself!*

But of course Pakhom is dead. The waiting Bashkirs click their tongues sympathetically, and they proceed to bury him.

They dig a hole for him. They measure it out with paces. They cut into the sod. The hole is six feet long and three feet wide. It is just the amount that poor Pakhom needs, no less and no more. It is just the amount that any man needs. Nice and dark and deep, and in area six feet by three feet, exactly.

With Great Thanks

In this endeavor, I stand on the shoulders of one memorable giant: the writer Andro Linklater, scion of a remarkable Scottish family, an author whose range of interests was breathtakingly broad, but whose later books, written just before his untimely death in 2013 at the age of sixty-eight, were devoted to matters of land measurement and ownership and, to me, were biblical in their authority. His *Owning the Earth* (Bloomsbury, 2013) is in particular a book for the ages, and was a profound inspiration for my very much more modest effort.

This book was born out of a discussion at home—my wife, Setsuko, and I were talking one morning about the iniquities of the then-current American immigration policy—concerning the degree to which land enclosure

in Britain might have played a part in the eighteenth-
and nineteenth-century transatlantic crossings. There
seemed a cruel irony about it: that those who found
themselves newly dispossessed on one side of the ocean
had then sailed across it and settled again, but only
promptly to dispossess those who already lived and
prospered in what to the migrant Britons was a new-
found and unowned country. The irony seemed worthy
of exploration, and of recounting.

Researching what then turned out to be a very
much more intricate and subtle story than I imagined
required a great deal of travel, from Latvia to New
Zealand, from the Scottish Highlands to the Ukrainian
steppe, to India and Japan and all by way of America's
midwestern prairies. Very many people generously took
my hand as I stumbled into these variously unfamiliar
territories—with my asking each time: *Who owns this
land and how did it come to be owned?* So here I list,
in alphabetical order, those who helped me in a variety of ways, offering their thoughts or their coffee or
their writings or their hospitality, and often all of these
things and more. I hope they will realize how vastly
grateful I am to them, and how this book was made so
much better as a consequence of their kindness, advice,
wisdom, and consideration. Any errors are mine and
mine alone.

So I owe my sincerest thanks to Kate Andrews, Fran Aramaki, Patricia Atkinson, Russell Baillie, Gill Baron, Marcy Bidney, Kenny Brown, Patricia Calhoun, Brett Chapman, Stephen Corry, Robert Cottrell, Robin Darwall-Smith, Uri Davis, Philip Deloria, Chris Dillon, Trent Duffy, Judge Caren Fox, David Freese, Donna Fujii, Charles Geisler, Peter Godwin, Jan deGraeve, Jenny Hansell, Robert Horneyold-Strickland, Jamie Howard, Wilson Isaac, Kristen Iversen, Ian Jack, Kirk Johnson, Miranda Johnson, Judy Joseph, Moira Kelly, Mariia Kravchenko, Barbara Lauriat, David Lazan, Robert Lee, Michael Levien, Rebecca Long, J T Moore, Willy van der Most, Ralph Nader, David Neiwert, Henk Pruntel, Hugh Raven, Wendy Reid, Jeanne Ryckmans, Salman Abu Sitta, Steve Small, Jim Smith, Jonathan Steffert, Joanna Storie, Mick Strack, Melanie Sturm, Toni Tack, Iain Thornber, Neal Ulevich, Paul Vercoe, Juliet Walker, Maggie Wells, Michael Wigan, Rick Wilcox, Michael Williams, Angus Winchester (not my son, though by coincidence I have one of that very name), and the indefatigable Rupert Winchester, who kindly helps me with all my books, and most assuredly is my son.

My two HarperCollins editors, Sara Nelson in New York and Arabella Pike in London, were enthused and supportive from the very start. On receipt of my initially untidy and unvarnished manuscript, they worked uncomplainingly to cut and polish and chamfer and finally to varnish the text to a high gleam: my gratitude to these fine practitioners of a noble craft is limitless. Likewise, I owe much to Mary Gaule, who took care of the mechanics of getting these words cast into cold type, or rendered into pixel or audio form, and who moreover researched and discovered and helped select the images and maps. Mary, in short, helped turn some hundreds of pages of dust-dry edited text into something agreeable to see, to have and to hold—and to read.

And not to forget those who so carefully engineered the arrangements with the publishers: WME's Suzanne Gluck, deservedly the best known of all New York's many agents, and in this instance backed and supported by her tirelessly helpful and cheerfully unflappable helpmeet, Andrea Blatt. It is worth remembering just when this book was written and was turned into the volume you now hold. The editing process began in April 2020, soon after the Great Pandemic got itself into high gear. Watching these agents and editors get to work reminded me of the familiar adage from Ginger Rogers, telling audiences that she did all that Fred Astaire did, only backward and in heels. My editors and agents, it turned out, did all that I had done for the book, only quite alone and isolated, and wearing masks. My thanks to them all is boundless.

And I mentioned that this book was born out of a breakfast-time discussion with my wife, Setsuko. She knows how grateful I am; but the extent of my gratitude is such that it is well worth saying once again to her: an almighty thank you!

A Glossary of Terms

Some Possibly Unfamiliar, Associated with Land and Its Ownership

ACRE. This ninth-century Germanic borrowing, which first signified the amount of land that could be ploughed by a team of oxen in a day, was by the thirteenth century legally defined as an area measuring 220 by 22 yards—4,840 square yards—and to be configured in any shape.

APPANAGE. A piece of territory—sometimes quite a large piece, as in all of Wales or Cornwall or Cumberland—set aside for the security for the younger children of a monarch or a prince.

BUNDLE OF RIGHTS. Central to the concept of ownership, at least in western society, are the five fundamental rights that are implicitly bundled with the title: the owner is free to possess the land; to

control it; to exclude others from it; to enjoy it; and
to dispose of it as is deemed fit.

CADASTRE. A public register of real property
assembled for the purpose of determining its value
and so how much tax should be levied.

CHENGBAO **SYSTEM.** An ancient Chinese arrangement
of tenancy—now much revived and corrupted—
whereby a farmer may have a thirty-year lease on a
tract of land and after supplying the state with an
agreed amount of foodstuffs may retain any surplus
for himself. The rules, laid down by the vaguely
formed local party collective, allowed for all manner
of venality, and have enriched many officials and
aided in the creation of vast cities clawed from the
Chinese countryside.

CIFTLIK. A system of land tenure in later Ottoman
times, by which Turkish military commanders seized
territory and ruled the local peasantry as serfs. It
replaced the ancient and rather more liberal Timar
system (qv).

CLEARANCE. Applied most commonly to Scottish
moorland, this often cruelly imposed Victorian-
era policy was designed to clear the countryside
of unprofitable crofters and replace them with
sheep, from which the landlords could expect
a more considerable and reliable income.

Victims of clearances often immigrated to North America.

CORNAGE. A form of rent determined by the number of horned cattle—*cornu* meaning horn in Latin—owned by the tenant.

COUVERTURE. A now technically obsolete U.S. legal doctrine that prevented women from owning real property. Prejudice against female landownership persists in some few states, with banks still reluctant to permit women to take out mortgages without a man as a cosigner.

CROFTING. The practice, peculiar in name to Scotland, of the peasantry having the right to farm some small portion of arable land, as tenancies. But in the nineteenth century the price of wool rose so steeply that major landowners reckoned it more profitable to raise sheep rather than rely on crofting for income, and so cleared the land of the resident crofters, with predictably melancholy consequences.

CUBIT. From the Latin for "elbow" this Roman and Egyptian unit of length measured from elbow to index fingertip—seventeen inches for the average citizen of Rome, up to four inches longer for the evidently sturdier residents of Cairo.

DAIMYO. A senior member of the Japanese nobility, answerable to the shogun, and the employer of a number of samurai as guardians of the lands and protectors of his family and his family's honor.

DEED. The formal written evidence, invariably a handsomely inscribed document of paper, parchment, or vellum, attesting to the ownership of the title to a piece of real estate, signed and sealed by an earlier owner.

DIGGERS. In 1649 the agrarian reformer Gerrard Winstanley led a group of radicals to dig up recently privatized land on St. George's Hill south of London. They also filled in ditches and tore down hedges with a view to giving the fields back to the common people. They called themselves True Levellers, but the press of the day called them The Diggers. Their movement failed.

DUNUM. An area of land, roughly equivalent to an English acre, used by the Ottoman administrators of their various conquered territories. The unit varied wildly: in Ottoman Iraq one dunum was equivalent to 2,500 square meters; in Palestine, little more than 900.

EMINENT DOMAIN. An ever-present reminder that the ultimate owner of land in many countries is a monarch. In Britain, where the reigning sovereign in theory has supervising authority over ownership of every acre in the country, there are occasions when a supposed owner is compelled to relinquish title to his lands, and for the supposed good of all. The British call the practice compulsory purchase, which is what it is. The American phrase, unnecessarily baroque, means essentially the same thing.

ENCLOSURE. The practice, which by the eighteenth century was generally backed by law, of fencing off areas of commonly used and hitherto unowned land and declaring it henceforward to be private property. The effects of enclosure on the development of society were profound, driving many dispossessed country dwellers to the cities, or else persuading them to emigrate.

***ENCOMENDADO* SYSTEM.** Land seized by Imperial Spain belonged, ipso facto, to the monarch. But the conquistadores who performed the seizing were allowed—*encomendado*—to use the local native people to farm and extract minerals from this land, and, as noble hidalgos, lived there and prospered, though without ownership.

ENFEOFF. To give a fief, or land that is offered in exchange for military service.

ESCHEAT. Land seized by the state for nonpayment of taxes—a reminder that ownership is invariably less than absolute, and can be forfeit, and that the state is almost everywhere the ultimate true owner of all.

ESTATE. The interest one has in land, which may vary from a tenancy to various forms of ownership. Real property refers to land and any immovable structures on it; personal property is the term used in law for the estate one has—usually ownership—in movable objects, from handbags to books to motor cars.

FALL. A seldom-used and somewhat elastic Scottish and northern English length measurement, equivalent to such older units as the rod, pole, and perch, and about one fortieth of a furlong, itself a measure that varies by region. Also, a measurement of volume in the marl-digging business.

FEDDAN. An areal unit of land measurement in some Arab countries and roughly equivalent to an English acre.

FEE SIMPLE. The term, derived from the French, denoting the absolute ownership of a piece of property, and its heritability to heirs and successors who then enjoy similar absolute ownership.

FEUDAL TENURE. Where one may, as a vassal, enjoy use of land in exchange for some obligation, usually military service, on behalf of the ultimate owner.

FOOT. A unit of linear measure, originally and roughly based on the length of an adult foot, but latterly codified against the standard yard measurement, of which one foot is precisely one third.

GEORGISM. Followers of the once wildly popular Victorian-era journalist and polemicist Henry George, who argued forcefully for the abolition of all taxes except one Land Value Tax, which would be assessed on the rental value of land.

GUNTER'S CHAIN. Invented by the English mathematician Edmund Gunter in 1623, this iron chain, consisting of one hundred links, and which at a specific temperature measures a length of exactly sixty-six feet, is used as the basis for much surveying, worldwide. Ten square chains is equivalent to one acre.

JAMES HARRINGTON. A seventeenth-century political theorist whose seminal book *Oceana* (1656) suggested limits on private landownership, since in his view those who owned too much land accreted to themselves an unhealthy amount of political power that would, ultimately, destroy them. His ideas were debated in the Rota Club, which he founded.

HEADRIGHT. A term of colonial settlement indicating the grant of landownership made in exchange for, say, the cost of transportation to a tract of unsettled territory—much of rural Virginia was given a fifty-acre headright exchange for the payment of passage fees.

HECTARE. An increasingly common metric measurement based on a square of one hundred meters. It is equivalent to 2.47 acres.

HOLODOMOR. The Ukrainian word meaning "to kill by starvation" and applied to the great famine of 1932 that resulted in between 3 million and 12 million deaths, and which many regard as a genocide perpetrated by the then–Soviet leader, Josef Stalin.

KIBBUTZ. A cooperative settlement, usually based on agriculture, deriving from an idea originating in the Jordan valley of Palestine in 1910, by Jewish settlers from Europe.

KULAK. A class of landholding peasantry of very moderate means, regarded with contempt by Stalin and subject to the most savage treatment during the Holodomor of the early 1930s.

LANDSCHAFT. The German classification of landscape is meticulous and detailed, and became bound up in the Nazi era with the so-called blood and soil movement with its racial and nationalist overtones. The word literally translates to *landscape*, but in this sense now extends well beyond into more sinister territory.

LEVITTOWN. William Levitt of New York built seven massive suburban housing developments—all houses factory-made on production lines, one every eleven

minutes—for soldiers returning from the Second
World War. Eminently affordable, and with lawns
and white picket fences, they seemed to represent
an ideal—but Levitt's company was criticized
for refusing to sell to Black people, and the word
Levittown soon became a byword for racial prejudice
and discrimination.

LI. A unit of measurement in China, originally
considered the length of extent of an average village,
or about one third of a western mile.

LIEN. The right—if documented as part of a contract
or agreement—to retain possession of another's
property, including land, for the failure to repay a
debt.

LYNCHET. An earth terrace, built up over the years
at the end of the lower slope of a ploughed field in
early southern England. If aligned in parallel these
banks of earth are known as strip lynchets and can
form early field boundaries between neighboring
farms.

KEN. A unit of measurement in Japan, standardized
today as the equivalent of 19/11 meter. Half a square
ken is the size of a tatami mat, another unit still
employed in the measurement of floor space.

EDDIE MABO. An indigenous Australian from the
Torres Strait Islands who fought for aboriginal land

rights, and won. The ruling in the case, *Mabo vs Queensland No. 2*, overturned the British notion of *terra nullius* and recognized aboriginal land rights for the first time. The ruling was handed down in June 1992, five months after Mabo died; he was much honored, but posthumously.

METES AND BOUNDS. The boundaries and limits of extent of some human-made settlement—a village or a town or a county.

METER (METRE). This unit of length was created in 1793, defined initially as one ten-millionth of the length of a quadrant of the earth measured from the North Pole to the Equator along the meridian that passes through Paris. Like all metric measurements it has been much refined, and is now the length of the path of light passing through a vacuum during a time interval of 1/299,792,458 of a second.

MILE. Originally a Roman unit of length, derived from the Latin *mille passus*, a thousand double-paces of a marching centurion, and about 1611 England yards. It has been codified in law since 1592 when Parliament in London declared a mile to be eight furlongs, or 5,280 feet, 1,760 yards.

MOU. A Chinese measurement of land area, defined today by customs agreements as 920.417 square yards. In the Tang dynasty, from the seventh to tenth centuries, every Chinese man was entitled to a loan of eighty mou of land, to be returned to the state on his death, and twenty mou that could be kept and inherited by his descendants.

PARASANG. An old Persian unit of length, some three or four miles. It was divided into thirty stadia.

PATENT. The ultimate document offering evidence of the ownership of land—an official government writ, signed and sealed by sovereign authority, showing to whom the land was first granted. All deeds and titles of today's ownership can in theory be traced back to the "letters patent" guaranteeing the rights of heritable possession to all in the subsequent chain of ownership.

PERCH. An old English measure of length, equal to 5.5 yards. Also a rod, pole, or lug. First employed in the fourteenth century.

PHYSIOCRACY. A form of natural-order government based on the understanding that since land is the only and ultimate true source of wealth, so the direct taxation of land should be the only true source of revenue. The idea first advanced by the French economist Francois Quesnay was later popularized by Henry George.

PICKSHAFT. The handle of a pick was briefly used as a length measure in fourteenth-century England.

PING. A traditional Chinese areal measurement unit, equal to some 3.3 square meters.

PLAT. A map, usually large-scale, of a building or an area of land on which buildings are, or are about to be, built.

PLETHRON. A unit of length in ancient Greece, or a square with one side measuring one plethron. There are one hundred Greek feet in a plethron, surprisingly almost exactly one hundred standard Imperial feet.

POLDER SYSTEM. Though this Dutch word strictly signifies the low-lying land reclaimed from the sea and kept dry by the use of dykes, its modern use now extends to mean the cooperative work needed to maintain the habitable nature of the land and of the country itself: the polder system suggests a national need to set politics aside and pull as one to offset harm.

PRIMOGENITURE. The custom and right by which the firstborn child—in most early instances the firstborn male child—inherits property or title to real estate.

PUBLIC LAND STATES. In those American states that originated after the original colonial holdings, all land belonged to the national government (though Native Americans naturally disagreed). Land patents were thus issued by the U.S. government's Public

Land Office and not, as in the earlier settlements, by the individual state governments.

PYONG. A Korean unit of areal measurement, a square kan, some 36 square Korean feet. Like the Chinese ping (qv), it is equal to some 3.33 square meters, or almost 4 square yards.

REDLINING. The practice—employed by government bodies, banks, insurance companies, and the like—of designating certain areas of a community as being of greater commercial risk, and so less eligible for loans or other services. Critics rightly regard redlining as racist and discriminatory.

ROOD. Confusingly, a rood was a historic English measure both of length (between eighteen and twenty-four feet) and area (a quarter of an acre). The word and area measurement are originally Dutch.

ROTA CLUB. Established by James Harrington (qv) during the political interregnum that followed the execution of King Charles I. The club, whose members included Christopher Wren and Samuel Pepys, debated utopian ideas, including radical approaches to land reform.

SACHEM. A term, most commonly used in the American northeast, for a Native American chief. A sachem of the Wampanoag in eastern Massachusetts is often credited for having saved the Pilgrim settlers from starvation.

SAMURAI. A soldier-retainer in service to a daimyo, a nobleman and landowner. Samurai, belonging to a revered military caste, usually carried two swords, and were ferociously skilled in using them.

SERF. Down at the very bottom of the feudal system, a serf worked as a laborer, essentially in no more than a modified form of slavery.

SHAKU. A traditional Japanese measure of length, the Japanese foot, allowed officially until 1966, still used for some cloth measurement today. One variety, the whale shaku, used baleen whiskers as ruler.

SHARECROPPER. A tenant farmer who works the land for its owner and receives a portion of the crop as compensation for doing so.

TERRA NULLIUS. The classically Imperial legal fiction claiming that sparsely populated lands unoccupied by (generally British) settlers belonged in law to nobody, and were thus susceptible to ownership by their European discoverers. Australia and much of interior North America were once regarded in this manner.

THEODOLITE. A barely portable instrument of lenses, compasses, and brass vernier scales and mounted usually on a tripod, combining great precision with cumbersome weight and size, used in surveying tracts of land, meridians, or entire countries.

TIMAR. An early form of land distribution during the Ottoman empire in which temporary and nonheritable grants of conquered territory were made to those cavalrymen, Janissaries, and even slaves who had taken part in the conquest.

TITLE. A legal right to the ownership of land or of other property, usually based on an aggregation of facts or evidence from the land's recorded history.

TORRENS TITLE. While serving as registrar general of South Australia in 1858, the Irish lawyer Sir Robert Torrens introduced a system of land registration that allowed the government to create certificates of ownership—Torrens titles—that decisively proved possession. The system has since been adopted near universally in Western nations.

TRESPASS. Defined in law as the wrongful and uninvited entry onto lands owned by another person, and the perpetration of damage, however trivial, to the owner's real property.

TSUBO. A Japanese unit of a measurement equivalent to two tatami mats.

USUFRUCT. The right to enjoy—but not to destroy, sell, or otherwise profoundly alter—land that belongs to someone else. Thomas Jefferson famously remarked that "the earth belongs—in usufruct—to the living."

VARA. The Spanish yard, generally equivalent to some thirty-three inches, and which is still occasionally found used as a measurement standard in Texas

YOJANA. A varying unit of distance employed in early fifth-century India and equating to between eight and nine miles.

YOKELET. In Kent, in southern England, a yokelet once signified a small arable farm that could be worked by one yoke of oxen. It became briefly used as a measure of land—somewhat larger than an acre, which originated in the size of a field that could be worked by just a single animal.

YOKING. A strip of ploughed land, usually no more than ten yards wide, fashioned by farmers in Scotland and far northern England.

Bibliography

Anderson, Sam. *Boom Town: The Fantastical Saga of Oklahoma City.* New York. Crown. 2018.

Applebaum, Anne. *Red Famine: Stalin's War on Ukraine.* London. Allen Lane. 2017.

Archives New Zealand. *The Treaty of Waitangi.* Wellington. Bridget Williams Books. 2017.

Baker, Alan R. and Gideon Biger (eds.). *Ideology and Landscape in Historical Perspective: Essays on the Meanings of Some Places in the Past.* Cambridge. Cambridge University Press. 1992.

Berkman, Richard L. and W. Kip Viscusi. *Damming the West: Ralph Nader's Study Group Report on the Bureau of Reclamation.* New York. Grossman. 1973.

Berry, Wendell. *The Gift of Good Land: Further Essays Cultural and Agricultural.* Berkeley, CA. Counterpoint. 1981.

Bowes, John P. *Land Too Good for Indians: Northern Indian Removal.* Norman. University of Oklahoma Press. 2016.

Brasher, Rex. *Secrets of the Friendly Woods.* New York. The Century Co. 1926.

Brewer, Richard. *Conservancy: The Land Trust Movement in America.* Hanover, NH. University Press of New England. 2003.

Brooke-Hitching, Edward. *The Phantom Atlas: The Greatest Myths, Lies and Blunders on Maps.* London. Simon & Schuster. 2016.

Butler, Jenna. *A Profession of Hope: Farming on the Edge of the Grizzly Trail.* Hamilton, ON. Wolsak and Wynn. 2015.

Butler, Samuel. *Erewhon, or Over the Range.* London. Trübner. 1872.

Byrnes, Giselle. *Boundary Markers: Land Surveying and the Colonisation of New Zealand.* Wellington. Bridget Williams Books. 2001.

Cahill, Kevin. *Who Owns the World: The Surprising Truth About Every Piece of Land on the Planet.* New York. Grand Central. 2010.

Calloway, Colin G. *The Indian World of George Wash-*

ington: The First President, the First Americans, and the Birth of the Nation. New York. Oxford University Press. 2018.

Christophers. Brett. *The New Enclosure: The Appropriation of Public Land in Neoliberal Britain*. London. Verso. 2018.

Conquest, Robert. *Harvest of Sorrow: Soviet Collectivization and the Terror-Famine*. New York. Oxford University Press. 1986.

Crane, Nicholas. *The Making of the British Landscape: From the Ice Age to the Present*. London. Weidenfeld & Nicolson. 2016.

Cronon, William. *Changes in the Land: Indians, Colonists, and the Ecology of New England*. New York. Hill & Wang. 1983.

Dartnell, Lewis. *Origins: How Earth's History Shaped Human History*. New York. Basic Books. 2019.

Dary, David. *Entrepreneurs of the Old West*. Lincoln. University of Nebraska Press. 1986.

Debo, Angie. *A History of the Indians of the United States*. Norman. University of Oklahoma Press. 1970.

Debo, Angie. *And Still the Waters Run: The Betrayal of the Five Civilized Tribes*. Princeton. Princeton University Press. 1940.

Denman, D. R. *Origins of Ownership: A Brief History of Land Ownership and Tenure in England from Earliest*

Times to the Modern Era. London. Allen & Unwin. 1958.

Devine, T. M. *The Scottish Clearances: A History of the Dispossessed.* London. Allen Lane. 2018.

Douglas, Roy. *Land, People and Politics: A History of the Land Question in the United Kingdom 1878–1952.* London. Allison & Busby. 1976.

Dressler, Camille. *Eigg: The Story of an Island.* Edinburgh. Polygon. 1998.

Dunbar-Ortiz, Roxanne. *An Indigenous Peoples' History of the United States.* Boston. Beacon Press. 2014.

Dunn, Shirley W. *The River Indians: Mohicans Making History.* Fleischmanns, NY. Purple Mountain Press. 2009.

Easdale, Nola. *Kairuri: The Measurer of Land.* Petone, New Zealand. Highgate. 1988.

Egan, Timothy. *The Worst Hard Time: The Untold Story of Those Who Survived the Great American Dust Bowl.* Boston. Houghton Mifflin. 2006.

Fairlie, Simon et al. (eds.). *The Land.* Bridport, Dorset. March 2006–present.

Fellmeth, Robert C. *Politics of Land: Ralph Nader's Study Group Report on Land Use in California.* New York. Grossman. 1973.

Ferguson, Niall. *Empire: How Britain Made the Modern World.* London. Allen Lane. 2013.

Ferrari, Marco, et al. *A Moving Border: Alpine Cartographies of Climate Change*. New York. Columbia Books on Architecture. 2018.

Foreman, Grant. *Indian Removal: The Emigration of Five Civilized Tribes of Indians*. Norman. University of Oklahoma Press. 1932.

Forster, E. M. *Abinger Harvest*. New York. Harcourt, Brace. 1936.

Gammage, Bill. *The Biggest Estate on Earth: How Aborigines Made Australia*. Crows Nest, NSW. Allen & Unwin. 2011.

Geisler, Charles C. (ed.). *Who Owns Appalachia?: Land Ownership and Its Impact*. Lexington, KY. University Press of Kentucky. 1981.

George, Henry. *Progress and Poverty: An Inquiry into the Cause of Industrial Depressions and of Increased Want with Increase of Wealth; the Remedy*. New York. Appleton & Co. 1879.

Godwin, Peter. *When a Crocodile Eats the Sun*. New York. Little, Brown. 2008.

Grandin, Greg. *The End of the Myth: From the Frontier to the Border Wall in the Mind of America*. New York. Henry Holt. 2019.

Greer, Allan. *Property and Dispossession: Natives, Empire and Land in Early Modern America*. Cambridge, UK. Cambridge University Press. 2018.

Griffiths, Billy. *Deep Time Dreaming: Uncovering Ancient Australia*. Carlton, VIC. Black Inc. 2018.

Hamsun, Knut. *Growth of the Soil*. (Orig: *Markens Grode*). New York. Knopf. 1921.

Hardy, Roger. *The Poisoned Well: Empire and Its Legacy in the Middle East*. New York. Oxford University Press. 2017.

Hewitt, Rachel. *Map of a Nation: A Biography of the Ordnance Survey*. London. Granta. 2010.

Heyman, Stephen. *The Planter of Modern Life: Louis Bromfield and the Seeds of a Food Revolution*. New York. Norton. 2020.

Hightower, Michael J. *1889: The Boomer Movement, the Land Run, and Early Oklahoma City*. Norman. University of Oklahoma Press. 2018.

Hinks, Arthur R. *Maps and Survey*. Cambridge. Cambridge University Press. 1913.

Hogue, Michel. *Metis and the Medicine Line: Crossing a Border and Dividing a People*. Chapel Hill. University of North Carolina Press. 2015.

Hoig, Stan. *The Oklahoma Land Rush of 1889*. Oklahoma City. Oklahoma Historical Society. 1989.

Hunter, James. *The Claim of Crofting: The Scottish Highlands and Islands, 1930–1990*. Edinburgh. Mainstream Publishing. 1991.

Hunter, James. *From the Low Tide of the Sea to the*

Highest Mountain Tops: Community Ownership of Land in the Highlands and Islands of Scotland. Isle of Lewis. The Islands Book Trust. 2012.

Hutchinson, Bruce. *The Struggle for the Border.* Don Mills, ONT. Oxford University Press. 1955.

Johnson, Miranda. *The Land Is Our History: Indigeneity, Law, and the Settler State.* New York. Oxford University Press. 2016.

Johnson, Richard B. (ed.). *History of Us: Nisenan Tribe of the Nevada City Rancheria.* Santa Rosa, CA. Comstock Bonanza Press. 2018.

Jorgensen, Neil. *A Guide to New England's Landscape.* Chester, CT. The Globe Pequot Press. 1977.

Kaplan, Robert D. *Earning the Rockies: How Geography Shapes America's Role in the World.* New York. Random House. 2017.

Keay, John. *The Great Arc: The Dramatic Tale of How India Was Mapped and Everest Was Named.* New York. HarperCollins. 2000.

King, Michael. *The Penguin History of New Zealand.* Auckland. Penguin. 2003.

King, Michael. *Whina: A Biography of Whina Cooper.* London. Hodder & Stoughton. 1983.

Kunstler, James Howard. *The Geography of Nowhere: The Rise and Decline of America's Man-Made Landscape.* New York. Touchstone. 1993.

Leopold, Aldo. *A Sand County Almanac: And Sketches Here and There*. New York. Oxford University Press. 1949.

Linklater, Andro. *Measuring America: How the United States Was Shaped by the Greatest Land Sale in History*. New York. Penguin. 2003.

Linklater, Andro. *Owning the Earth: The Transforming History of Land Ownership*. London. Bloomsbury. 2013.

Lopez, Barry (ed.). *Home Ground: Language for an American Landscape*. San Antonio, TX. Trinity University Press. 2006.

Lynam, Edward (ed.). *The Mapmaker's Art: Essays on the History of Maps*. London. The Batchworth Press. 1953.

Mabey, Richard. *The Common Ground: a Place for Nature in Britain's Future?* London. Hutchinson. 1980.

Maier, Charles S. *Once Within Borders: Territories of Power, Wealth and Belonging since 1500*. Cambridge, MA. Harvard University Press. 2016.

Marsden, Philip. *Rising Ground: A Search for the Spirit of Place*. London. Granta. 2014.

Marsden, Philip. *The Summer Isles: A Voyage of the Imagination*. London. Granta. 2019.

Marshall, James M. *Land Fever: Dispossession and the*

Frontier Myth. Lexington, KY. University Press of Kentucky. 1986.

McGuire, Lloyd H., Jr. *Birth of Guthrie: Oklahoma's Run of 1889 and Life in Guthrie in 1889 and the 1890s*. San Diego. Privately Published. 1998.

Mingay, G. E. *Parliamentary Enclosure in England: An Introduction to Its Causes, Incidence and Impact 1750–1850*. Harlow, Essex. Longman. 1997.

Ministry of Justice, New Zealand. *150 Years of the Maori Land Court*. Wellington. New Zealand Government. 2015.

Mitchell, John Hanson. *Ceremonial Time: Fifteen Thousand Years on One Square Mile*. Cambridge, MA. Perseus Books. 1984.

Mitchell, John Hanson. *Trespassing: An Inquiry into the Private Ownership of Land*. Reading, MA. Perseus Books.1998.

Monbiot, George. *Feral: Rewilding the Land, the Sea, and Human Life*. London. University of Chicago Press. 2014.

Moss, Graham. *Britain's Wasting Acres: Land Use in a Changing Society*. London. The Architectural Press. 1981.

Neiwert, David A. *Strawberry Days: How Internment Destroyed a Japanese American Community*. New York. Palgrave Macmillan. 2005.

Nicolson, I. F. *The Mystery of Crichel Down.* Oxford. Clarendon Press. 1986.

Nikoliç, Zoran. *The Atlas of Unusual Borders.* Glasgow. HarperCollins, 2019.

O'Donnell, Edward T. *Henry George and the Crisis of Inequality: Progress and Poverty in the Gilded Age.* New York. Columbia University Press. 2015.

O'Malley, Vincent. *The Great War for New Zealand: Waikato 1800–2000.* Wellington. Bridget Williams Books. 2019.

O'Malley, Vincent. *The New Zealand Wars.* Wellington. Bridget Williams Books. 2019.

O'Malley, Vincent et al. *The Treaty of Waitangi Companion: Maori and Pakeha from Tasman to Today.* Auckland. Auckland University Press. 2010.

Osborn, William C. *The Paper Plantation: Ralph Nader's Study Group on the Pulp and Paper Industry in Maine.* New York. Grossman. 1974.

Polhemus, John and Richard Polhemus. *Up on Preston Mountain: The Story of an American Ghost Town.* Fleischmanns, New York. Purple Mountain Press. 2005.

Prebble, John. *The Highland Clearances.* London. Secker & Warburg. 1963.

Purdy, Jedediah. *This Land Is Our Land: The Struggle*

for a New Commonwealth. Princeton. Princeton University Press. 2019.

Quinn, Tom. *The Reluctant Billionaire: The Tragic Life of Gerald Grosvenor, 6th Duke of Westminster.* London. Biteback Publishing. 2018.

Rackham, Oliver. *The History of the Countryside.* London. Dent. 1986.

Rader, Andrew. *Beyond the Known: How Exploration Created the Modern World and Will Take Us to the Stars.* New York. Scribner. 2019.

Rees, Tony. *Arc of the Medicine Line: Mapping the World's Longest Undefended Border Across the Western Plains.* Lincoln. University of Nebraska Press. 2007.

Reeves, Richard. *Infamy: The Shocking Story of the Japanese American Internment in World War Two.* New York. Henry Holt. 2015.

Robillard, Walter G. and Donald A. Wilson. *Brown's Boundary Control and Legal Principles.* Hoboken. Wiley. 2003.

Rothstein, Richard. *The Color of Law: A Forgotten History of How Our Government Segregated America.* New York. Liveright. 2017.

Rowse, Tim. *Indigenous and Other Australians Since 1901.* Sydney. University of New South Wales Press. 2017.

Sapp, Rick. *Native Americans State by State*. New York. Quarto. 2018.

Schama, Simon. *The Story of the Jews*. London. Harper-Collins. 2013.

Schulten, Susan. *A History of America in 100 Maps*. London. The British Library. 2018.

Segeren, W. G. (ed.). *Polders of the World*. Arnhem. Published Proceedings of Symposium, Lelystad. 1982.

Shorto, Russell. *Amsterdam: A History of the World's Most Liberal City*. New York. Random House. 2013.

Shrubsole, Guy. *Who Owns England? How We Lost Our Green and Pleasant Land and How to Take It Back*. London. William Collins. 2019.

Stamp, L. Dudley and W. G. Hoskins. *The Common Land of England and Wales*. London. Collins. 1963.

Stannard, David E. *American Holocaust: The Conquest of the New World*. New York. Oxford University Press. 1992.

Stein, Mark. *How the States Got Their Shapes*. New York. HarperCollins. 2008.

Tolstoy, Leo. *How Much Land Does a Man Need? And Other Stories*. London. Penguin. 1994.

Toynbee, Arnold J. *A Study of History*. London. Oxford University Press. 1946.

Tree, Isabella. *Wilding: Returning Nature to Our Farm*. London. Picador. 2018.

Treuer, Anton. *Atlas of Indian Nations.* Washington, D.C. National Geographic. 2013.

Treuer, David. *The Heartbeat of Wounded Knee: Native America from 1890 to the Present.* New York. Riverhead Books. 2019.

Trudolyubov, Maxim. *The Tragedy of Property: Private Life, Ownership and the Russian State.* Cambridge. Polity Press. 2018.

Vitek, William and Wes Jackson (eds.). *Rooted in the Land: Essays on Community and Place.* New Haven. Yale University Press. 1996.

Watson, Peter. *Ideas: A History of Thought and Invention, from Fire to Freud.* New York. HarperCollins. 2005.

Weaver, John C. *The Great Land Rush and the Making of the Modern World, 1650–1900.* Montreal. McGill-Queen's University Press. 2003.

Weisman, Alan. *The World Without Us.* New York. St. Martin's Press. 2007.

Wigan, Michael. *The Salmon: The Extraordinary Story of the King of Fish.* London. William Collins. 2013.

Wightman, Andy. *The Poor Had No Lawyers: Who Owns Scotland (and How They Got It).* Edinburgh. Birlinn. 2015.

About the Author

SIMON WINCHESTER is the acclaimed author of many books, including *The Professor and the Madman*, *The Men Who United the States*, *The Map That Changed the World*, *The Man Who Loved China*, *A Crack in the Edge of the World*, and *Krakatoa*, all of which were *New York Times* bestsellers and appeared on numerous best and notable lists. In 2006, Winchester was made an officer of the Order of the British Empire (OBE) by Her Majesty the Queen. He resides in western Massachusetts.